D0066304

WORKBOOK TO ACCOMPANY

CARPENTRY

FOURTH EDITION

FLOYD VOGT

THOMSON

DELMAR LEARNING

Australia ■ Canada ■ Mexico ■ Singapore ■ Spain ■ United Kingdom ■ United States

THOMSON

DELMAR LEARNING

Workbook to Accompany Carpentry, 4th Edition

Floyd Vogt

Vice President, Technology and Trades ABU:
David Garza

Director of Learning Solutions:
Sandy Clark

Acquisitions Editor:
Alison Weintraub

Product Manager:
Jennifer A. Thompson

Channel Manager:
William Lawrensen

Marketing Coordinator:
Mark Pierro

Production Director:
Mary Ellen Black

Production Manager:
Andrew Crouth

Senior Project Editor:
Christopher Chien

Art & Design Specialist:
Mary Beth Vought

Technology Project Manager:
Kevin Smith

Technology Project Specialist:
Linda Verde

Editorial Assistant:
Maria Conto

COPYRIGHT © 2006 by Delmar Learning. Thomson, the Star Logo, and Delmar Learning are trademarks used herein under license.

Printed in the United States of America
1 2 3 4 5 XX 07 06

For more information contact
Thomson Delmar Learning
Executive Woods
5 Maxwell Drive, PO Box 8007
Clifton Park, NY 12065-8007
Or find us on the World Wide Web at
www.delmarlearning.com

ALL RIGHTS RESERVED. No part of this work covered by the copyright hereon may be reproduced or used in any form or by any means—graphic, electronic, or mechanical, including photocopying, recording, taping, Web distribution or information storage and retrieval systems—without written permission of the publisher.

For permission to use material from this text or product, contact us by
Tel   (800) 730-2214
Fax   (800) 730-2215
www.thomsonrights.com

Library of Congress Cataloging-in-Publication Data:
Card Number: 2005048589
ISBN 1-4018-7070-8

NOTICE TO THE READER

Publisher does not warrant or guarantee any of the products described herein or perform any independent analysis in connection with any of the product information contained herein. Publisher does not assume, and expressly disclaims, any obligation to obtain and include information other than that provided to it by the manufacturer.

The reader is expressly warned to consider and adopt all safety precautions that might be indicated by the activities herein and to avoid all potential hazards. By following the instructions contained herein, the reader willingly assumes all risks in connection with such instructions.

The Publisher makes no representation or warranties of any kind, including but not limited to, the warranties of fitness for particular purpose or merchantability, nor are any such representations implied with respect to the material set forth herein, and the publisher takes no responsibility with respect to such material. The publisher shall not be liable for any special, consequential, or exemplary damages resulting, in whole or part, from the readers' use of, or reliance upon, this material.

Contents

SECTION TWO

Rough Carpentry 91

Preface

This workbook is designed to accompany *Carpentry, 4th Edition,* and is intended to provide you, the student, with a wide variety of activities to reinforce the important topics introduced in your textbook. Each chapter includes a set of corresponding questions and exercises that will help you successfully accomplish the course content, including the following:

- *Multiple Choice Questions* highlight key concepts and help you prepare for quizzes and exams.
- *Completion Questions* allow you to practice learning key terms and definitions for communicating on the jobsite.
- *Identification Exercises* help you to appropriately identify components of wood products and carpentry procedures.
- A *Math Problem-Solving* in each chapter provides word problems containing various situations where math skills are critical to the accurate completion of a job.

- *Sketching Exercises* provide an opportunity for you to practice identifying key print symbols.
- *Discussion Questions* get you thinking! Potential scenarios are presented to encourage you to practice your creativity and problem-solving skills.
- *Four Building for Success Exercises* accompany each section of chapters and units and provide practical advice toward developing key initiatives and traits for advancing in the construction industry – the promotion of safety, effective communication, solid teamwork, and presenting quality workmanship.

After reading each chapter in *Carpentry,* it is advisable to practice the questions and exercises included in the corresponding chapter. If need be, refer back to your *Carpentry* text until you are confident that you have mastered the material. Remember, practice make perfect!

1 Wood

Multiple Choice

Write the letter for the best answer on the line next to the number of the sentence.

_____ 1. The carpenter must understand the nature and characteristics of wood to
_____ .

    A. protect it from decay
    B. select it for the appropriate use
    C. work it with the proper tools
    D. all of the above

_____ 2. The insulating value of 1″ of wood is the equivalent of _____ of brick.

    A. 6″
    B. 10″
    C. 14″
    D. 18″

_____ 3. _____ is a wood that is known for its elasticity.

    A. Oak
    B. Maple
    C. Pine
    D. Hickory

_____ 4. The natural substance that holds wood's many hollow cells together is called
_____ .

    A. pith
    B. cambium layer
    C. lignin
    D. sapwood

_____ 5. Tree growth takes place in the _____ .

    A. heartwood
    B. medullary rays
    C. pith
    D. cambium layer

_____ 6. The central part of the tree that is usually darker in color is called the _____ .

    A. sapwood
    B. heartwood
    C. springwood
    D. medullary rays

_____ 7. Wood growth that is rapid and takes place in the _____ is usually light in color and
rather porous.

    A. spring
    B. summer
    C. fall
    D. winter

_____ 8. Periods of fast or slow growth can be determined by _____ of the tree.
  A. counting the annual rings
  B. measuring the height
  C. studying the width of the annual rings
  D. measuring the circumference

_____ 9. _____ is an example of a hardwood that is softer than some softwoods.
  A. Basswood
  B. Oak
  C. Redwood
  D. Cherry

_____ 10. All softwoods are _____ .
  A. close-grained
  B. cone-bearing
  C. open-grained
  D. A and B

Completion

Complete each sentence by inserting the best answer on the line near the number.

_____ 1. _____ trees lose their leaves once a year.

_____ 2. Softwoods come from _____ trees, commonly known as evergreens.

_____ 3. Water passes upward through the tree in the _____ .

_____ 4. Wood that comes from deciduous trees is classified as _____ .

_____ 5. Fir comes from the _____ classification of wood.

_____ 6. Oak is an example of _____ -grained wood.

_____ 7. The _____ of cedar, cypress, and redwood are extremely resistant to decay.

_____ 8. Open-grained lumber has large _____ that show tiny openings or pores in the surface.

_____ 9. Cedar can always be identified by its characteristic _____ .

_____ 10. The best way to learn the different types of wood is by _____ with them.

Identification: Cross-section of Wood

Identify each term, and write the letter of the best answer on the line next to each number.

_____ 1. pith

_____ 2. sapwood

_____ 3. cambium layer

_____ 4. medullary rays

_____ 5. heartwood

_____ 6. annual rings

_____ 7. bark

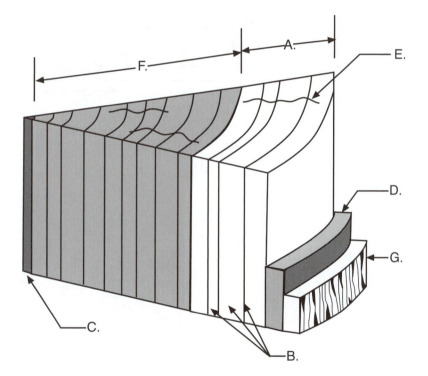

Math Problem-Solving

Solve the following math problems.

_____ 1. A logger signs a contract with a homeowner to cut trees from her property. If 17 ash, 36 cherry, 21 fir, 45 hemlock, 75 maple, and 3 oak trees are cut, what is the total number of trees?

_____ 2. What is the total number of hardwood logs to be cut?

_____ 3. What is the total number of softwood to be cut?

_____ 4. Add the whole numbers 246, 1350, 78, and 9.

_____ 5. What is the length of measurement A in figure shown below?

_____ 6. What is the length of measurement B in figure shown below?

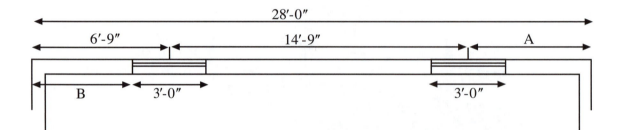

2 Lumber

Multiple Choice

Write the letter for the best answer on the line next to the number of the sentence.

_____ 1. The process of restacking lumber in a way that allows air to circulate uses pieces known as _____.
   A. blocking
   B. spacers
   C. stickers
   D. stackers

_____ 2. The best appearing side of a piece of lumber is its _____ side.
   A. face
   B. visage
   C. veneer
   D. select

_____ 3. Most logs are sawed using the _____ method.
   A. plain-sawed
   B. quarter-sawed
   C. edge-grained
   D. a combination of the plain and quarter-sawed

_____ 4. Cracked ceilings, sticking doors, squeaking floors, and many other problems can occur from using _____ lumber.
   A. recycled
   B. green
   C. seasoned
   D. quarter-sawed

_____ 5. The moisture content of lumber is expressed as a percentage _____ .
   A. of its total weight
   B. of its total volume
   C. of the weight of its free water
   D. of weight to volume

_____ 6. Wood has reached its _____ when all of the free water is gone.
   A. equilibrium moisture content
   B. stabilization point
   C. fiber-saturation point
   D. dehydration point

_____ 7. Lumber that is under 2″ thick has the classification of _____ .
   A. timbers
   B. boards
   C. dimensional
   D. joists

_____ 8. Dimension lumber is in the following category: _____
    A. under 2″ thick
    B. 2″-4″ thick
    C. 5″ and thicker
    D. open-grained only

_____ 9. The best grade of hardwood as established by the National Hardwood Association is
_____ .
    A. select
    B. first and seconds
    C. No. 1 commons
    D. choice

_____ 10. Parallel cracks between the annual rings in wood that are sometimes caused by
storm damage are known as _____ .
    A. shakes
    B. crooks
    C. checks
    D. cups

Completion

Complete each sentence by inserting the best answer on the line near the number.

_____ 1. _____ -sawed lumber is the least expensive method of sawing.

_____ 2. _____ -sawed lumber is less likely to warp or shrink.

_____ 3. The _____ uses a great amount of skill in determining the most
efficient and conservative way to cut a log.

_____ 4. When lumber is first cut from the log it is called _____ lumber.

_____ 5. The heavy weight of green lumber is due to its high _____ content.

_____ 6. The low form of plant life that causes wood to decay is known as
_____ .

_____ 7. Wood with a moisture content of below _____ percent will not decay.

_____ 8. Lumber used for framing should not have a moisture content over
_____ percent.

_____ 9. Lumber used for interior finish should not have a moisture content over
_____ percent.

_____ 10. _____ moisture content occurs when the moisture content of the lumber is the same as the surrounding air.

_____ 11. S4S means the lumber was surfaced on _____ sides.

_____ 12. Crooks, bows, twists, and cups are classified as _____ .

Identification: Cut Lumber

Identify each term, and write the letter of the best answer on the line next to each number.

_____ 1. crook

_____ 2. quarter-sawed   A.

_____ 3. twist

_____ 4. cup   B.

_____ 5. check   C.

_____ 6. bow

_____ 7. plain-sawed

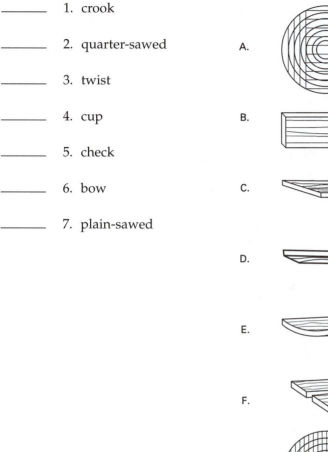

D.

E.

F.

G.

Math Problem-Solving

Solve the following math problems.

_____ 1. If one 2 × 6-10' board weighs 35 pounds, how many pounds will 50 boards weigh?

_____ 2. One person can stack and sticker 222 boards in one hour. How long will it take to stack and sticker 3330 boards?

_____ 3. How many board feet of wood are there in 4 boards that are 1″ × 6″ × 12′ long?

_____ 4. How many board feet of wood is there in 750-2 × 10-16′?

_____ 5. How many 1 × 12-10′ boards are there in 1000 board feet?

Discussion

Write your answer(s) on the lines below.

1. Describe the difference between air dried and kiln dried lumber.

    _____

    _____

    _____

2. Describe the difference between nominal and actual dimensions.

    _____

    _____

    _____

3. Describe some of the factors one must keep in mind when properly storing lumber on the job site.

    _____

    _____

    _____

    _____

3 Rated Plywood and Panels

Multiple Choice

Write the letter for the best answer on the line next to the number of the sentence.

_____ 1. A/an _____ is a very thin layer of wood.
- A. underlayment
- B. com-ply
- C. span
- D. veneer

_____ 2. With the use of engineered panels _____ .
- A. construction progresses faster
- B. more surface protection is provided than with solid lumber
- C. lumber resources are more efficiently used
- D. all of the above

_____ 3. Cross-graining in the manufacture of plywood refers to _____ .
- A. touch sanding the grain
- B. the use of open-grained hardwoods
- C. the grain of each succesive layer is at a right angle to the next one
- D. the placement of the peeler log on the lath

_____ 4. The American Plywood Association is concerned with quality supervision and testing of _____ .
- A. waferboards
- B. composites
- C. oriented strand board
- D. all of the above

_____ 5. The letters A, B, C, and D indicate _____ .
- A. span rating
- B. exposure durability classification
- C. the quality of the panel veneers
- D. strength grades

_____ 6. Douglas fir and southern pine are classified in the _____ strength grade.
- A. plugged C
- B. group 1
- C. 303
- D. 32/16

_____ 7. A performance rated panel meets the requirements of the _____ .
- A. panel's end use
- B. sawyer
- C. EPA
- D. United States Forest Service

_____ 8. The left-hand number in a span rating denotes the maximum recommended suport spacing when the panel is used for _____.

    A. roof sheathing
    B. subflooring
    C. siding
    D. underlayment

Completion

Complete each sentence by inserting the best answer on the line near the number.

_____ 1. The _____ is the largest trade association that tests the quality of plywood and other engineered panels.

_____ 2. The sheets of veneer that are bonded together to form plywood are also known as _____ .

_____ 3. Specially selected logs mounted on a huge lathe are known as _____ logs.

_____ 4. The highest appearance quality of a panel veneer is designated by the letter _____.

_____ 5. Panels with a _____ grade or better are always sanded smooth.

_____ 6. V-groove, channel groove, striated, brushed, and rough-sawed are all special surfaces used in the manufacture of _____ .

_____ 7. Most panels manufactured with oriented strands or wafers are known as _____.

_____ 8. Exposure durability of a panel is located on the _____ .

Matching

Write the letter for the best answer on the line near the number to which it corresponds.

_____ 1. veneer

_____ 2. span rating

_____ 3. exposure 1

_____ 4. exposure 2

_____ 5. exterior

_____ 6. plywood

_____ 7. oriented strand board

_____ 8. grade stamp

A. may be exposed to weather during moderate delays

B. wood fibers arranged in layers at right angles

C. cross-laminated, layered plies glued and bonded under pressure

D. may be exposed to weather during long delays

E. term used to describe the layers or plies of engineered panel

F. appears as two numbers separated by a /

G. may be permanently exposed to weather or moisture

H. assures the product has met quality and performance requirements

Identification: Label Information

Identify each term, and write the letter of the best answer on the line next to each number.

_____ 1. thickness

_____ 2. mill number

_____ 3. panel grade

_____ 4. national research board report number

_____ 5. exposure durability classification

_____ 6. span rating

APA

A. ———————→ **RATED SHEATHING**

B. ———————→ **32/16** 1/2 INCH ←——————— D.

SIZED FOR SPACING

C. ———————→ **EXPOSURE 2**

000 ←——————— E.

NRB-108 ←——————— F.

Math Problem-Solving

Solve the following math problems.

_____ 1. Each sheet of plywood measures 4′ × 8′. How many square feet will 24 sheets cover?

_____ 2. What is the average thickness of a ply in a piece of ½″ plywood if it is constructed with 4 plies?

_____ 3. One board weighs 2 pounds per foot and a ¾-ton truck is able to carry 1500 pounds. How many boards 12′ long can the truck carry?

_____ 4. What percent is 7 of 55?

_____ 5. What is the percent moisture content if 3 ounces of water is removed from a wood block with a dry weight of 14 ounces?

4 Nonstructural Panels

Multiple Choice

Write the letter for the best answer on the line next to the number of the sentence.

_____   1. Non-structural particleboard is used in the construction industry for _____ .
   - A. cabinet construction
   - B. kitchen countertops
   - C. the core of veneer doors
   - D. all of the above

_____   2. The best choice of plywood to be installed as a painted soffit is _____ .
   - A. A-A
   - B. A-B
   - C. A-C
   - D. C-D

_____   3. The highest quality particleboard _____ .
   - A. contains the same size particles throughout
   - B. is 100% sawdust
   - C. has large wood flakes in the center with the particle size decreasing the closer to the surface
   - D. usually has a rough surface texture

_____   4. High density fiberboards are called _____ .
   - A. particleboard
   - B. softboard
   - C. oriented strand board
   - D. hardboard

_____   5. *Masonite* is a brand name for _____ .
   - A. softboard
   - B. duraflake
   - C. hardboard
   - D. particleboard

_____   6. _____ is a brand name for softboard.
   - A. *Celotex*
   - B. *Fibrepine*
   - C. *Masonite*
   - D. all of the above

_____   7. To protect exterior softboard wall sheathing from moisture during construction it is impregnated with _____ .
   - A. lignin
   - B. asphalt
   - C. oil
   - D. creosote

_____ 8. _____ is a well-known reference on products used in the construction industries.
- A. The Dodge Report
- B. Dunn and Bradstreet
- C. Sweets Architectural File
- D. Thompson's Products

Completion

Complete each sentence by inserting the best answer on the line near the number.

_____ 1. _____ plywood is used where appearance is important on both sides.

_____ 2. The quality of _____ is indicated by its density per cubic foot.

_____ 3. _____ are manufactured as high density, medium density, and low density boards.

_____ 4. Lignin is utilized in the manufacturing process of _____ .

_____ 5. Tempered hardboard panels are coated with _____ and baked.

_____ 6. Most hardboard producers belong to the _____ .

_____ 7. Decorative ceiling panels for suspended ceilings are often made from _____ .

Math Problem-Solving

Solve the following math problems.

_____ 1. What is the exponent in the phrase $7^3 + 14$?

_____ 2. What is the value of 16^2?

_____ 3. What is the value of 6^4?

_____ 4. What is the square root of 144?

_____ 5. Find the square of 15.

_____ 6. Find the cube of 3.

5 Laminated Veneer Lumber

Multiple Choice

Write the letter for the best answer on the line next to the number of the sentence.

_____   1. During the manufacture of laminated veneer lumber, the grain in each layer of veneer is placed _____ to the previous one.
   A. parallel
   B. at right angles
   C. diagonally
   D. at 22.5°

_____   2. Laminated veneer lumber was first used to make _____ .
   A. automobile frames
   B. airplane propellers
   C. underlayment
   D. canoes

_____   3. The first commercially produced laminated veneer lumber for building construction was patented as _____ .
   A. *Micro-Lam*
   B. *Gang-Lam*
   C. *Struclam*
   D. *Versa-Lam*

_____   4. Laminated veneer lumber uses _____ in its construction.
   A. basswood
   B. poplar and aspen
   C. Douglas fir or southern pine
   D. mostly hardwoods

_____   5. Laminated veneer lumber is widely used in _____ .
   A. wood frame construction
   B. subflooring
   C. exterior siding
   D. interior paneling

_____   6. A typical 1¾″ beam of laminated veneer lumber contains _____ layers of veneer.
   A. 3–5
   B. 5–8
   C. 10–15
   D. 15–20

_____   7. The thickness of LVL veneers ranges from _____ of an inch.
   A. ⅟₃₂–⅛
   B. ⅟₁₀–³⁄₁₆
   C. ¼–⅜
   D. ⁵⁄₁₆–⁷⁄₁₆

_____ 8. Laminated veneer lumber is suited for _____ .
  A. load-carrying beams over window and door openings
  B. scaffold planks
  C. concrete forming
  D. all of the above

_____ 9. During its manufacture, LVL veneers are _____ .
  A. expanded
  B. densified
  C. coated with oil
  D. bonded with an interior adhesive

_____ 10. The usual thickness of LVL is _____ .
  A. ¾″
  B. 1″
  C. 2″ and 4″
  D. 1½″ and 1¾″

Math Problem-Solving

Solve the following math problems.

_____ 1. What is the total price of 16 sheets of MDF if each costs $23.95?

_____ 2. What is the tax that would be added to a $1724.23 purchase if the tax rate is 8%?

_____ 3. How many sheets of particle board will be needed to build a countertop measuring 3′ × 16′?

_____ 4. What is the total sale price of $525.00 of items with 8% sales tax?

_____ 5. High density particle board has a density of 55 pounds per cubic foot. What would be the density of 8 sheets of the same material?

6 Parallel Strand and Laminated Strand Lumber

Multiple Choice

Write the letter for the best answer on the line next to the number of the sentence.

_____ 1. Parallel strand lumber provides the building industry with _____ .
   A. sheet materials
   B. small dimension lumber
   C. large dimension lumber
   D. all of the above

_____ 2. Parallel strand lumber meets an environmental concern by using _____ .
   A. strands not made of wood
   B. small diameter second growth trees
   C. only old-growth trees
   D. mostly hardwoods

_____ 3. Parallel strand lumber is manufactured _____ .
   A. using a process that is identical to plywood
   B. much the same as particleboard
   C. using a microwave pressing process
   D. using ultrasound to bond the strands

_____ 4. Parallel strand lumber is a material _____ .
   A. that is available in 4′ × 8′ sheets
   B. that sometimes contains defects like knots and shakes
   C. that can be used wherever there is a need for a large beam or post
   D. whose demand will decrease in the future

_____ 5. In comparison to solid lumber, parallel strand lumber _____ .
   A. is consistent in strength throughout its length
   B. has few differences
   C. lengths are not as long
   D. may contain checks

_____ 6. The registered brand name for laminated strand lumber is _____ .
   A. *Lam-stran*
   B. *Parallam*
   C. *TimberStrand*
   D. *Struclam*

_____ 7. Laminated strand lumber presently is made from _____ .
   A. Douglas fir only
   B. longer strands than those used in parallel strand lumber
   C. surplus over-mature aspen trees
   D. southern pine

_____ 8. Laminated strand lumber _____ .
   A. is made of wood strands under 12″ long
   B. is not designed to carry heavy loads
   C. can be made from very small logs
   D. all of the above

Math Problem-Solving

Provide the best answer for each of the following questions.

1. The length of line A is _____ inches.

2. The length of line B is _____ inches.

3. The length of line C is _____ inches.

4. The length of line D is _____ inches.

5. The length of line E is _____ inches.

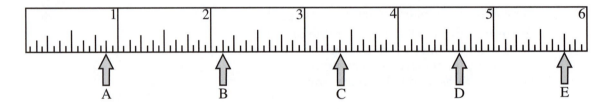

7 Wood I-Joists

Multiple Choice

Write the letter for the best answer on the line next to the number of the sentence.

_____ 1. Wood I-joists get their name from _____ .
   A. the woman who invented them
   B. the fact that they are long in length
   C. the pattern in which they are installed
   D. their I shape

_____ 2. The webs of wood I-joists may be made of _____ .
   A. wire mesh
   B. rebar
   C. oriented strand board
   D. hardboard

_____ 3. Wood I-joists are used for _____ .
   A. floor joists and roof rafters
   B. studs
   C. posts
   D. girders

_____ 4. The flanges of a wood I-beam are often made of _____ .
   A. LVL
   B. PSL
   C. LSL
   D. glulams

_____ 5. Flanges of a solid wood I-beam are joined together with _____ .
   A. finger joints
   B. cold joints
   C. screws
   D. rivets

_____ 6. The wood I-joist was invented in _____ .
   A. 1969
   B. 1972
   C. 1975
   D. 1983

Completion

Complete each sentence by inserting the best answer on the line near the number.

_____ 1. The "I" shape utilized in a wood I-beam increases its _____ .

_____ 2. The _____ of a wood I-beam may be made from laminated veneer
lumber or specially selected finger-jointed solid wood lumber.

_____ 3. Plywood, laminated veneer lumber, or oriented strand board may be used in the _____ of the beam.

_____ 4. The manufacturing process of wood I-beams consists of _____ top and bottom flanges to a web.

_____ 5. Wood I-beams are produced to approximate _____ moisture content.

_____ 6. Wood I-beams are available in depths from _____ to _____ inches.

_____ 7. Beams with larger webs and flanges are designed to carry _____ loads.

_____ 8. Wood I-beams are available in lengths of up to _____ feet long.

Math Problem-Solving

Provide the best answer for each of the following questions.

1. The length of line A is _____ inches.

2. The length of line B is _____ inches.

3. The length of line C is _____ inches.

4. The length of line D is _____ inches.

5. The length of line E is _____ inches.

6. The length of line F is _____ inches.

7. The length of line G is _____ inches.

8. The length of line H is _____ inches.

9. The length of line I is _____ inches.

10. The length of line J is _____ inches.

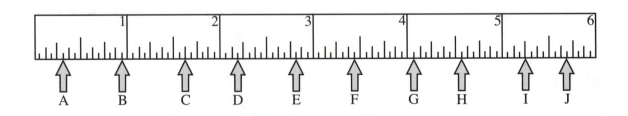

8 Glued-Laminated Lumber

Multiple Choice

Write the letter for the best answer on the line next to the number of the sentence.

_____ 1. The force that tends to decrease the length of a member is _____ .
  A. compression
  B. sheer
  C. tension
  D. push in

_____ 2. Lam layup refers to the sequence of lam grades from _____ .
  A. left to right
  B. right to left
  C. bottom to top
  D. end to end

_____ 3. Glulam beams must always be installed with the _____ .
  A. "BOTTOM" stamp towards the ground
  B. "TOP" stamp toward the sky
  C. "END" stamp towards the supports
  D. APA stamp facing down

_____ 4. ANSI stands for _____ .
  A. American National Standards Institute
  B. American National Stress Institute
  C. Association of National Stress Indicators
  D. Association of National Stress Institute

_____ 5. Premium appearance grade glulams are rated for _____ .
  A. warehouses
  B. architectural value
  C. strength
  D. both strength and critical appearance

_____ 6. The grade of glulam most likely to be used in a shopping mall where cost and appearance are equally important would be _____ .
  A. architectural appearance
  B. industrial appearance
  C. premium appearance
  D. select appearance

Completion

Complete each sentence by inserting the best answer on the line near the number.

_____ 1. Glued laminated lumber is commonly called _____ .

_____ 2. Glued laminated beams and joists are _____ than natural wood of the same size.

_____ 3. Aside from structural strength, glued laminated beams are _____ as well.

_____ 4. The individual pieces of stock in glued laminated lumber are known as _____ .

_____ 5. When a load is imposed on a glulam beam that is supported on both ends, the topmost lams are said to be in _____ .

_____ 6. _____ is a force applied to a member that tends to increase its length.

_____ 7. The lower stressed sections of glued laminated lumber are located in the beam's _____ .

_____ 8. The sequence of lam grades from the bottom to top of glued laminated lumber is referred to as _____.

_____ 9. Glued laminated beams come with one edge stamped _____ .

_____ 10. Glued laminated lumber is manufactured in three different grades for _____ .

_____ 11. Glued laminated beams are usually available in lengths from _____ to _____ feet in 2 foot increments.

_____ 12. The widths of glued laminated beams range from _____ to _____ inches.

Discussion

Write your answer(s) on the lines below.

1. What are the advantages of using engineered lumber over solid lumber?

   _____

   _____

2. Over time, will the engineered lumber products industry help or harm the lumber industry?

   _____

   _____

Math Problem-Solving

Solve the following math problems.

_____ 1. How many inches are in an LVL that is 4'-10" long?

_____ 2. How many feet and inches in a PSL that is 183"?

_____ 3. How many yards in 67'?

_____ 4. How many inches in 16 yards?

_____ 5. How many inches in an LSL that is 3 yards, 2 feet, 25 inches long?

9 Nails, Screws, and Bolts

Multiple Choice

Write the letter for the best answer on the line next to the number of the sentence.

_____ 1. Nails are made with many different kinds of materials. Which type of material below is not used to make nails?
   A. stainless steel
   B. brass
   C. masonry
   D. copper

_____ 2. The actual sizes of a 6d, 10d, and 16d nails are _____ respectively.
   A. 1½″, 2½″ and 3½″
   B. 1½″, 3″ and 3½″
   C. 2″, 3″ and 4″
   D. 2″, 3″ and 3½″

_____ 3. The best nail to assemble temporary structures such as concrete formwork is a _____ nail.
   A. masonry
   B. duplex
   C. common
   D. galvanized

_____ 4. Small finishing nails are referred to as _____ .
   A. finish nails
   B. casing nails
   C. brads
   D. box nails

_____ 5. The best choice of headed nail for nailing close to the edge of thinner is

   _____ .
   A. box nail
   B. casing nail
   C. finish nail
   D. common nail

_____ 6. Screws are identified by their _____ .
   A. length
   B. gauge
   C. head
   D. all of the above

_____ 7. The best fastener for joining thin sheets of metal is _____ .
   A. lag screws
   B. Phillips screws
   C. self-drilling screws
   D. drywall screws

_____ 8. The type of fastener designed to be secured with a wrench or socket set is _____ .
   A. lag screw
   B. machine bolt
   C. carriage bolt
   D. all of the above

_____ 9. The fastener with an oval head over a squared section is called
   A. machine bolt
   B. carriage bolt
   C. stove bolt
   D. all of the above

_____ 10. Drywall screws are designed _____ .
   A. with bugle shaped heads
   B. with sharp pointed tips
   C. to be driven into wood and thin metal
   D. all of the above

Completion

Complete each sentence by inserting the best answer on the line near the number.

_____ 1. Steel nails that are uncoated are called _____ nails.

_____ 2. Steel nails that are coated with zinc to prevent rusting are called _____ .

_____ 3. _____ galvanized nails have a heavier coating than electroplated galvanized.

_____ 4. Moisture reacting with two different types of metal causes _____ , which, in time, results in disintegration of one of the metals.

_____ 5. The diameter or thickness of a nail is referred to as its _____ .

_____ 6. Wedge-shaped nails that are stamped from thin sheets of metal are known as _____ nails.

_____ 7. The preferred nail for fastening exterior finish is the _____ nail.

_____ 8. Small finishing nails sized according to gauge and length in inches are known as _____ .

_____ 9. To prevent them from bending, masonry nails are made from _____ steel.

_____ 10. The pointed end of the screw is called the _____ point.

_____ 11. Steel screws that have no coating are known as _____ screws.

_____ 12. _____ screws are larger than wood screws and are turned with a wrench instead of a screwdriver.

_____ 13. The square section under the oval head of a carriage bolt prevents the bolt from _____ as the nut is being turned.

_____ 14. The proper tool used to turn a stove bolt is a _____ .

Identification: Nails, Screws, and Bolts

Identify each term, and write the letter of the best answer on the line next to each number.

_____ 1. duplex nail

_____ 2. finish nail

_____ 3. casing nail

_____ 4. roofing nail

_____ 5. common nail

_____ 6. gimlet point

_____ 7. threads

_____ 8. shank

_____ 9. lag screw

_____ 10. carriage bolt

_____ 11. machine bolt

_____ 12. stove bolt

_____ 13. staple

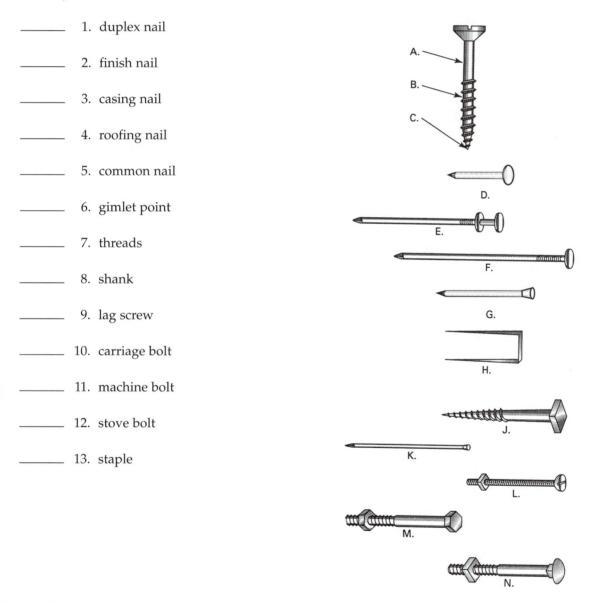

Math Problem-Solving

Solve the following math problems.

_____ 1. What is the sum of 6⅜", 5¼" and 10⅞"?

_____ 2. Subtract 11¾" from 20⅝".

_____ 3. Convert ³⁄₁₆" to a decimal.

_____ 4. Convert 0.68" to the nearest sixteenth of an inch.

_____ 5. What is one-third of 14'-8"? Answer in terms of inches to the nearest sixteenth.

10 Anchors and Adhesives

Multiple Choice

Write the letter for the best answer on the line next to the number of the sentence.

_____ 1. A _____ is a heavy duty anchor.
   A. wedge anchor
   B. expansion anchor
   C. nylon nail anchor
   D. concrete screw

_____ 2. Holes may be drilled directly in the masonry through the mounting holes of the fixture being installed if the _____ is used.
   A. toggle bolt
   B. drop-in anchor
   C. stud anchor
   D. conical screw

_____ 3. The concrete screw is a _____ type of anchor.
   A. heavy duty
   B. medium duty
   C. light duty
   D. all of the above

_____ 4. Lead and plastic anchors are also called _____ .
   A. split-fast anchors
   B. lag shields
   C. hollow wall fasteners
   D. inserts

_____ 5. When using chemical anchoring systems, it is important to _____ .
   A. thoroughly clear the hole of all dust
   B. properly torque the bolt into the system
   C. immediately stress test the bond
   D. all of the above

_____ 6. A disadvantage of using toggle bolts is that _____ .
   A. their use is limited to solid walls
   B. if removed the toggle falls off inside the wall
   C. the diameter of the hole has to be the same as the bolt
   D. hole depth is critical

_____ 7. A common name for a hollow wall expansion anchor is _____ .
   A. universal plug
   B. conical screw
   C. self-drilling anchor
   D. molly screw

_____ 8. Conical screws are used on _____ .
   A. gypsum board
   B. cement block
   C. strand board
   D. A and C

_____ 9. Joist hangers are a form of _____ .
   A. universal anchor
   B. wood to wood connector
   C. hollow wall connector
   D. wood to concrete connector

_____ 10. Polyvinyl acetate is a _____ glue.
   A. yellow
   B. mastic
   C. contact
   D. white

_____ 11. Contact cement is widely used for _____ .
   A. applying plastic laminates on countertops
   B. interior trim
   C. exterior finish
   D. framing

_____ 12. _____ is a moisture-resistant glue.
   A. Yellow
   B. White
   C. Plastic resin
   D. Aliphatic resin

_____ 13. _____ is a type of mastic that may be used in cold weather, even on wet or frozen wood.
   A. Urea resin
   B. Resorcinol resin
   C. Contact cement
   D. Construction adhesive

_____ 14. When applying troweled mastics it is important to _____ .
   A. apply heavily
   B. be sure the depth and spacing of trowel's notches are correct
   C. thoroughly brush as well as trowel the mastic
   D. mix the proper ratio of hardener to the mastic

Identification: Anchors, Bolt, and Screw

Identify each term, and write the letter of the best answer on the line next to each number.

_____ 1. joist hanger

_____ 2. stud anchor

_____ 3. sleeve anchor

_____ 4. drop-in anchor

_____ 5. lag shield

_____ 6. nylon nail anchor

_____ 7. toggle bolt

_____ 8. expansion anchor

_____ 9. conical screw

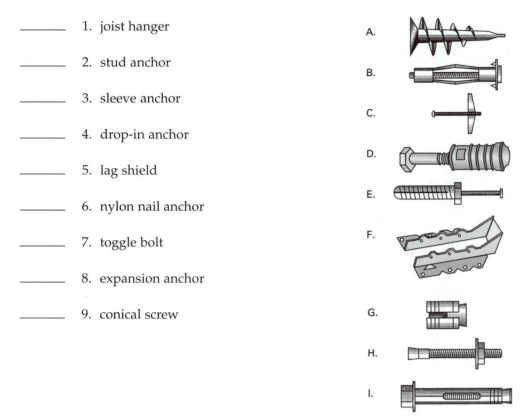

A.

B.

C.

D.

E.

F.

G.

H.

I.

Math Problem-Solving

Solve the following math problems.

_____ 1. Multiply ⅓ × ¾ and express answer as a fraction in lowest terms.

_____ 2. Divide ⅚ by ⅔ and express answer as a fraction in lowest terms.

_____ 3. How many pounds will 72 joist hangers weigh if each weighs 7½ ounces?

_____ 4. How much will 5000 drywall screws cost if each screw is priced at $0.005?

_____ 5. Express ⅝ as a decimal.

11 Layout Tools

Multiple Choice

Write the letter for the best answer on the line next to the number of the sentence.

_____ 1. Early in carpenter training it is important that _____ is mastered.
   A. the essex board foot table
   B. quick and accurate measuring
   C. finger gauging
   D. the octagon scale

_____ 2. Most of the rules and tapes used by the carpenter have clearly marked increments of
   _____ .
   A. yards
   B. 32nds of an inch
   C. 16ths of an inch
   D. metric conversions

_____ 3. Pocket tapes are available in _____ lengths.
   A. 50 and 100 foot
   B. 35 and 50 foot
   C. 6 to 35 foot
   D. all of the above

_____ 4. Red and black highlights on a tape measure are located at every ____ inches.
   A. 12
   B. 16
   C. 19.2
   D. all of the above

_____ 5. The combination square functions as _____ .
   A. a depth gauge
   B. a layout or test device for 90° and 45° angles
   C. a marking gauge
   D. all of the above

_____ 6. The rafter layout tool that can double as a guide for a portable circular saw is the
   _____ .
   A. speed square
   B. combination square
   C. trammel point
   D. framing square

_____ 7. The side of the framing square known as the face is the one that has the _____ on it.
   A. essex board foot table
   B. brace table
   C. hundredths scale
   D. manufacturer's name stamped

_____ 8. The most often used table on the framing square is the _____ .
   A. rafter table
   B. brace table
   C. essex board foot table
   D. hundredths scale

_____ 9. The _____ is used to lay out or test angles other than those laid out with squares.
   A. trammel point
   B. butt gauge
   C. folding ruler
   D. sliding T-bevel

_____ 10. In the absence of trammel points, the same type of layout can be made with _____ .
   A. a plumb bob
   B. a thin strip of wood with a brad through it for a center point
   C. butt markers
   D. a line level

Completion

Complete each sentence by inserting the best answer on the line near the number.

_____ 1. The longer of the two legs of a framing square is known as the _____ .

_____ 2. In construction the term level is used to indicate that which is _____ .

_____ 3. The term _____ is used to mean the same as vertical.

_____ 4. The pair of tubes located in the center of the level is used to determine _____ .

_____ 5. When reading a level where the bubble is touching the right side line of the vial, the _____ side should be raised to make the surface level.

_____ 6. When using the line level, it is important that the level is placed as close to the _____ of the line as possible.

_____ 7. Although very accurate indoors, the _____ can be difficult to use outside when the wind is blowing.

_____ 8. _____ is the technique of laying out stock to fit against an irregular surface.

Matching

Write the letter for the best answer on the line near the number to which it corresponds.

_____ 1. folding rule

_____ 2. pocket tape

_____ 3. framing square

_____ 4. plumb bob

_____ 5. chalk line

_____ 6. butt marker

_____ 7. line level

_____ 8. wing dividers

A. gives only an approximate levelness; is not accurate

B. practically useless when wet

C. joints must be oiled occasionally

D. used to mark hinge gains

E. is often called a scriber

F. hook slides back and forth slightly

G. is suspended from a line

H. outside corner is called the heel

Measuring

Provide the best answer for each of the following questions.

1. The length of line A is _____ inches.

2. The length of line B is _____ inches.

3. The length of line C is _____ inches.

4. The length of line D is _____ inches.

5. The length of line E is _____ inches.

6. The length of line F is _____ inches.

7. The length of line G is _____ inches.

8. The length of line H is _____ inches.

9. The length of line I is _____ inches.

10. The length of line J is _____ inches.

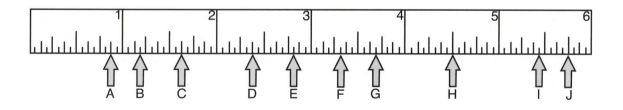

12 Boring and Cutting Tools

Multiple Choice

Write the letter for the best answer on the line next to the number of the sentence.

_____ 1. A technique to keep wood from splitting out the bottom when drilling with an auger bit is to _____.
   A. drill from both sides of the stock
   B. use high bit turning speed
   C. drill with increasing pressure on the bit
   D. all of the above

_____ 2. Firmer chisels are used mostly on _____ .
   A. heavy framing
   B. millwork
   C. finish work
   D. locksets

_____ 3. The longest bench plane is a _____ plane.
   A. jointer
   B. jack
   C. smooth
   D. block

_____ 4. A bevel that is referred to as hollow ground is one with a surface that is _____ .
   A. convex
   B. concave
   C. flat
   D. raised

_____ 5. When reshaping the bevel on a wood chisel by grinding, it is important to _____ .
   A. use safety goggles
   B. cool the blade by dipping it in water frequently
   C. maintain a width to the bevel that is approximately twice the thickness of the blade
   D. all of the above

_____ 6. Chisels and plane irons need to be ground _____ .
   A. every time they become dull
   B. before each time they are used
   C. when the bevel has lost its hollow ground shape
   D. to reestablish a new wire-edge on the tool's back

_____ 7. The handsaw that is designed to cut with the grain is the _____ .
   A. crosscut saw
   B. compass saw
   C. hacksaw
   D. ripsaw

_____ 8. Stock is handsawn with the face side up because _____ .
    A. the action of the saw will splinter the bottom side
    B. it prevents the saw from jumping at the start of the cut
    C. it keeps the saw from binding in the cut
    D. the motion of the saw is built up before it hits the bottom of the board

_____ 9. The _____ is used to make circular cuts in wood.
    A. ripsaw
    B. hacksaw
    C. compass saw
    D. miterbox

_____ 10. The _____ crosscut saw is used to cut across the grain of trim and finished boards.
    A. 11- or 12-point
    B. 5½-point
    C. 7- or 8-point
    D. 4½-point

Completion

Complete each sentence by inserting the best answer on the line near the number.

_____ 1. The term _____ is often used to denote cutting larger holes in wood.

_____ 2. The diameter of the circle made by the handle of a bit brace is known as its _____ .

_____ 3. As an auger bit is turned, the _____ score the circle in advance of the cutting lips.

_____ 4. The number of sixteenths of an inch in the diameter of an auger bit designates its _____ .

_____ 5. The _____ is usually used to bore holes over 1″ in diameter.

_____ 6. The _____ forms a recess for a flat head screw to set flush with the surface of the material in which it is driven.

_____ 7. The _____ is used to cut recesses in wood for such things as door hinges and locksets.

_____ 8. _____ come in several sizes and are used for smoothing rough surfaces and to bring work down to size.

_____ 9. The _____ is a small plane designed to be held in one hand.

_____ 10. A/An _____ is used to shape the bevel on a wood chisel or plane iron.

_____ 11. To produce a keen edge on a chisel it must be whetted using a/an _____ .

_____ 12. When whetting the flat side of a chisel or plane iron, always hold the tool _____ on the stone.

_____ 13. _____ are generally used to cut straight lines on thin metal.

_____ 14. Handsaws used to cut across the grain of lumber are called _____ saws.

_____ 15. The _____ saw is used to make circular cuts in wood.

_____ 16. Coping and hacksaw blades are usually installed with the teeth pointing _____ from the blade.

_____ 17. The proper hacksaw blade for fast cutting in thick metal is the _____ -toothed blade.

_____ 18. A saw similar to the compass saw, but designed especially for gypsum board, is the _____ saw.

_____ 19. The _____ is used to cut angles of various degrees in finish lumber.

_____ 20. Ripsaws have teeth shaped like rows of tiny _____ .

Math Problem-Solving

Use a rafter square to solve the following problems.

_____ 1. What is the length of the longest side of the framing square?

_____ 2. What is the width of the tongue?

_____ 3. What is the first row number in the Essex Board Foot Scale under the 12?

_____ 4. In the Rafter Table, what is the first row number under the 8″ mark?

_____ 5. In the Rafter Table, what is the last row number under the 10″ mark?

Discussion

Write your answer(s) on the lines below.

1. In a time when power tools are available, is it still important that carpenters know how to choose and skillfully use hand tools?

_____

_____

2. Why is it important for the carpenter to purchase good quality tools?

_____

_____

_____

13 Fastening and Dismantling Tools

Multiple Choice

Write the letter for the best answer on the line next to the number of the sentence.

_____ 1. The 22-ounce hammer is most popularly used on _____ .
- A. finish work
- B. rough work
- C. general work
- D. all of the above

_____ 2. The tip of a nail set is _____ to prevent it from slipping off the nail head.
- A. flattened
- B. convexed
- C. concave
- D. pointed

_____ 3. A hammer should be held _____ .
- A. firmly and close to the end of the handle
- B. firmly and in the middle of the handle
- C. in different places on the handle depending on the nails you are using
- D. loosely and in the middle

_____ 4. Toenailing is a technique of driving nails _____ .
- A. at an angle to fasten the end of one piece to another
- B. overhead in hard-to-reach places
- C. straight in to hardwood
- D. sideways in to end grain

_____ 5. To prevent wood from splitting or the nail from bending in hardwood, _____ .
- A. presoak the wood
- B. angle the nail
- C. use hardened nails
- D. drill a hole slightly smaller than the nail shank

_____ 6. The higher the number of the point size of a Phillips screwdriver, the _____ .
- A. higher the quality of the steel
- B. deeper the depth of the cross shaped tip
- C. larger the diameter of the point
- D. lower the steel's quality

_____ 7. If a screw has a flat head _____ .
- A. the shank hole must be countersunk
- B. a bit brace must be used to drive the screw
- C. a spiral screwdriver must be used
- D. a washer must be used under the head

_____ 8. The _____ is a dismantling tool available in lengths from 12″–36″ that is used to withdraw spikes and for prying purposes.
   A. nail claw
   B. pry bar
   C. wrecking bar
   D. cats paw

_____ 9. C-clamp sizes are designated by _____ .
   A. their overall length
   B. the sweep of their handle
   C. their throat opening
   D. the outside length of the C

_____ 10. The proper smoothing tool to use when a considerable amount of stock is to be removed is the _____ .
   A. flat file
   B. half-round file
   C. triangular file
   D. rasp

Completion

Complete each sentence by inserting the best answer on the line near the number.

_____ 1. The most popular hammer for general use is the _____ .

_____ 2. As a general rule, use nails that are _____ longer than the thickness of the material being fastened.

_____ 3. To prevent the filler from falling out, it is important that finish nails are set at least _____ deep.

_____ 4. Blunting or cutting off the point of a nail helps to prevent _____ the wood.

_____ 5. In preparation for driving a screw, a shank hole and a _____ hole must be drilled.

_____ 6. When drilling the shank hole, use a _____ to prevent drilling too deep.

_____ 7. If the material to be fastened is thick, the screw may be set below the surface by _____ to gain additional penetration without resorting to a longer screw.

_____ 8. The nail claw is commonly called a _____ .

_____ 9. To turn nuts, lag screws, bolts, and other objects, a/an _____ is often used.

Math Problem-Solving

Solve the following math problems.

_____   1.   What is the measured distance, to the nearest eighth, between the outside corners of the tongue and blade of a framing square?

_____   2.   What is the perimeter of a square with one side measuring 5″?

_____   3.   What is the perimeter of a rectangle that measures 15′ × 35′?

_____   4.   What is true about the diagonals of rectangles?

_____   5.   If the threads of a wood screw clamp are spaced 16 threads per inch, how many turns will the screw need to be turned to open the jaws 1¾″?

14 Saws, Drills, and Drivers

Multiple Choice

Write the letter for the best answer on the line next to the number of the sentence.

_____ 1. The _____ is the most used portable power tool the carpenter uses.
   A. saber saw
   B. electric circular saw
   C. heavy-duty drill
   D. reciprocating saw

_____ 2. The size of a portable circular saw is determined by _____ .
   A. its horsepower rating
   B. the length of the saw's base
   C. the saw's amperage
   D. the blade diameter

_____ 3. When operating the portable circular saw, _____ .
   A. keep the saw clear of the body until the blade has completely stopped
   B. follow the layout line closely
   C. be sure the stock's waste is over the end of the supports, not between them
   D. all of the above

_____ 4. Splintering occurs along the _____ with the portable circular saw.
   A. layout line of the stock
   B. opposite side of the stock from the saw
   C. beginning of the cut
   D. bottom of the kerf

_____ 5. When using the portable circular saw, the easiest way to prevent splintering is to _____ .
   A. lay out and cut on the finish side of the stock
   B. score the layout line with a sharp knife before cutting
   C. lay out and cut on the side opposite the finish side
   D. B and C

_____ 6. A saber saw blade with more teeth per inch will give _____ .
   A. faster, but rougher it cuts
   B. slower, but smoother it cuts
   C. the stroke range of saw
   D. plunge cuts

_____ 7. The reciprocating saw is primarily used for _____ .
   A. cutting straight lines
   B. finish cuts
   C. bevel and miter cuts
   D. roughing-in

8. The _____ saw would be the carpenters' most likely choice for cutting roof sheathing.
   A. circular
   B. saber
   C. reciprocating
   D. bayonet

9. _____ are more efficent when used in a hammer drill.
   A. Spade bits
   B. Hole saws
   C. Twist drills
   D. Masonry bits

10. The size of a portable power drill is determined by the _____ .
   A. maximum opening of its chuck
   B. maximum revolutions per minute it can achieve
   C. horsepower of its motor
   D. maximum torque rating it has received

Completion

Complete each sentence by inserting the best answer on the line near the number.

1. To prevent fatal accidents from shock, make sure the tool is plugged into a _____ outlet.

2. It is important to use the proper size _____ to prevent excessive voltage drop.

3. _____ circular saw blades stay sharper longer than high speed steel blades.

4. When following a layout line with a circular saw, any deviation from the line can cause the saw to bind and possibly _____ .

5. To cut out the hole for a sink in a countertop, it is necessary to make a _____ cut.

6. Saber saws cut on the _____ .

7. For fast cutting in wood, the reciprocating saw should be set in the _____ mode.

8. Drills that have a D-shaped handle generally are classified as _____ duty.

9. A disadvantage of a _____ is that a hole cannot be made partially through the stock.

10. Holes in metal must always be _____ to prevent the drill from wandering off center.

Matching

Write the letter for the best answer on the line near the number to which it corresponds.

_____ 1. GFCI

_____ 2. electric circular saw

_____ 3. saber saw

_____ 4. reciprocating saw

_____ 5. light-duty drills

_____ 6. heavy-duty drills

_____ 7. twist drills

_____ 8. spade bit

A. sometimes called a jigsaw

B. ¼″ or ⅜″ chuck capacity

C. used for boring holes in rough work

D. makes small holes in wood or metal

E. ½″ chuck capacity

F. will trip at 5 milliamperes

G. sometimes called a sawzall

H. commonly called the skilsaw

Math Problem-Solving

Solve the following math problems.

_____ 1. Add the measurements 6′-4″ + 2′-8″ + 5′-7″. Answer in terms of feet-inches.

_____ 2. What is the area of a square with a side measuring 14″?

_____ 3. What is the area of a rectangle that measures 30′ × 48′?

_____ 4. What is the area of a triangle with a base of 32′ and a height of 8′?

_____ 5. What is the area of a floor that measures 6′-6″ by 8′-8″? Answer in decimal form rounded to two decimal places.

Discussion

Write your answer(s) on the lines below.

1. What attitude should an operator of a portable power tool have?

_____

_____

2. List some general safety rules that apply to all portable power tools.

_____

_____

15 Planes, Routers, and Sanders

Multiple Choice

Write the letter for the best answer on the line next to the number of the sentence.

_____ 1. The tool used to cut a chamfer in stock is ____.
   A. hand plane
   B. power plane
   C. router
   D. all of the above

_____ 2. The power jointer plane can take up to _____ off the stock with one pass.
   A. ⅛″
   B. ¼″
   C. ¹⁄₁₆″
   D. ⁵⁄₁₆″

_____ 3. The part of the router that holds the bit is known as the _____ .
   A. cutter head
   B. template
   C. pilot guide
   D. chuck

_____ 4. Always advance the router into the stock in a direction that is _____ .
   A. clockwise on outside edges and ends
   B. counterclockwise when making internal cuts
   C. against the rotation of the bit
   D. with the rotation of the bit

_____ 5. The tool most often used to cut material for countertops is ____.
   A. laminate trimmer
   B. power plane
   C. belt sander
   D. finish sander

_____ 6. Extreme care must be taken while sanding when a _____ is applied.
   A. painted coating
   B. transparent coating
   C. penetrating stain
   D. all of the above

_____ 7. The adjusting screw on the front of a belt sander is used to _____ .
   A. retract the front roller when changing belts
   B. track the belt and center it on the roller
   C. control the belt's speed
   D. retract the belt's guard

_____ 8. When operating the belt sander, it is important to always _____ .
   A. exert downward pressure on the sander
   B. sand against the grain
   C. tilt the sander in several different directions
   D. keep the electrical cord clear of the tool

_____ 9. A major disadvantage of operating the finish sander in the orbital mode is that _____ .
   A. it is slower than the oscillating mode
   B. it leaves scratches across the grain
   C. it is more likely to overheat the sander
   D. the belt needs to be tracked frequently

_____ 10. Commonly used grits for finish sanding are _____ .
   A. 100 or 120
   B. 60 or 80
   C. 30 or 36
   D. 12 or 16

Completion

Complete each sentence by inserting the best answer on the line near the number.

_____ 1. The power _____ is used to smoothe and straighten long edges, such as when fitting doors to openings.

_____ 2. The motor on a power plane turns a _____ .

_____ 3. The power _____ plane is ordinarily operated with one hand.

_____ 4. The _____ plane is similar to a power block plane but has a heavy-duty sleeve instead of a cutter head.

_____ 5. The _____ is used to make many different cuts such as grooves, dadoes, rabbets, and dovetails.

_____ 6. At least ½″ of the router bit must be inserted into the router's _____ .

_____ 7. Caution must always be exercised when operating the router, because the _____ is unguarded.

_____ 8. Improper use of the _____ has probably ruined more work than any other tool.

_____ 9. The most widely used sandpaper on wood is coated with _____ .

_____ 10. Sandpaper _____ refers to the size of the abrasive particles.

Math Problem-Solving

Solve the following math problems.

_____ 1. What is the average of 75, 85, and 98?

_____ 2. What is the area of a triangle with a base of 9'-3" and a height of 4'?
Answer in terms of square feet.

_____ 3. What is the value of the area of a trapezoid A, if A = ½(15 + 37)18?

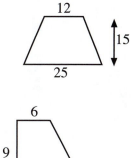

_____ 4. What is the area of the trapezoid shown?

_____ 5. What is the area of the cross section shown?

16 Fastening Tools

Multiple Choice

Write the letter for the best answer on the line next to the number of the sentence.

_____ 1. Pneumatic fastening tools are powered by _____ .
   A. explosive powder cartridges
   B. mechanical leverage
   C. compressed air
   D. disposable fuel cells

_____ 2. A _____ would be used to fasten subfloor.
   A. finish nailer
   B. brad nailer
   C. roofing stapler
   D. light duty framing gun

_____ 3. Nails come _____ for easy insertion into the framing gun's magazine.
   A. glued in strips
   B. attached end to end
   C. in preloaded plastic cassettes
   D. individually packed

_____ 4. Cordless nailing guns eliminate the need for _____ .
   A. air compressors
   B. long lengths of air hoses
   C. extra set-up time
   D. all of the above

_____ 5. A battery and spark plug is found in _____ nailing guns.
   A. pneumatic
   B. cordless
   C. powder-actuated
   D. mechanical leverage

_____ 6. The fuel cell in a cordless framing nailer can deliver enough energy to drive about _____ nails.
   A. 1,200
   B. 2,500
   C. 3,200
   D. 4,000

_____ 7. When the nailing gun's trigger is depressed, a fastener is _____ .
   A. immediately driven
   B. driven when the gun's nose touches the work
   C. repeatedly ejected from the gun
   D. automatically actuated

_____ 8. Many states require certification to operate _____ .

     A. pneumatic nailers
     B. cordless nailers
     C. powder-actuated drivers
     D. mechanical staplers

_____ 9. The strength of a powder charge can be determined by its _____ .

     A. color code
     B. diameter
     C. number
     D. length

_____ 10. In selecting a powder charge, always use _____ .

     A. a charge that has more than enough power
     B. the weakest charge that will do the job
     C. a charge that will penetrate the drive pin below the surface
     D. A and C

Completion

Complete each sentence by inserting the best answer on the line near the number.

_____ 1. A nail set is not required and the possibility of marring the wood is avoided when using the _____ nailer.

_____ 2. Nails for roof nailers come in _____ of 120, which are easily loaded in a nail canister.

_____ 3. Air compressors are needed to operate _____ nailers.

_____ 4. The _____ nailer is used to fasten small moldings and trim.

_____ 5. Within the cordless nailer is a/an _____ engine that forces a piston down to drive the fastener.

_____ 6. A cordless finish nailer will drive about _____ nails before the battery needs to be charged.

_____ 7. A nailing gun can operate only when the work contact element is firmly _____ against the work and the trigger pulled.

_____ 8. Never leave an unattended gun with the _____ attached.

_____ 9. Specially designed pins are driven into masonry and steel by _____ drivers.

_____ 10. Never pry a powder charge out with a screwdriver or knife, as this could result in an _____ .

Math Problem-Solving

Solve the following math problems.

_____ 1. What is the perimeter of the figure A?

_____ 2. What is the area of the
              figure A?

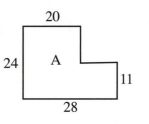

_____ 3. What is the perimeter of the figure B?

_____ 4. What is the area of the
              figure B?

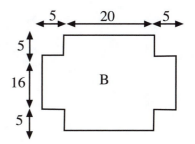

_____ 5. What is the area of the 5′ wide sidewalk shown in
              figure C?

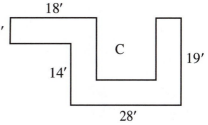

17 Circular Saw Blades

Multiple Choice

Write the letter for the best answer on the line next to the number of the sentence.

_____ 1. The more teeth on a circular saw blade, the _____ cut.
   A. faster it can
   B. rougher the surface of the
   C. smoother the surface of the
   D. longer between sharpening it can

_____ 2. If a blade is allowed to overheat, it is likely to _____ .
   A. loose its shape and wobble at high speed
   B. bind in the cut
   C. possibly cause a kickback
   D. all of the above

_____ 3. Coarse-tooth blades are suited for cutting _____ .
   A. thin, dry material
   B. heavy, rough lumber
   C. plastic laminated material
   D. where the quality of the cut surface is important

_____ 4. Every tooth on the _____ blade is filed or ground at right angles to the face of the blade.
   A. crosscut saw
   B. ripsaw
   C. combination
   D. A and C

_____ 5. The sides of all the teeth on a _____ blade are alternately filed or ground on a bevel.
   A. crosscut saw
   B. ripsaw
   C. combination
   D. B and C

_____ 6. The _____ is a feature of a circular saw blade that allows the blade to expand during normal operation.
   A. gullet
   B. hook angle
   C. arbor
   D. all of the above

_____ 7. The ideal carbide-tipped blade for cutting plastic laminated material is the

   _____ .

   A. triple chip grind
   B. alternative top bevel
   C. square grind
   D. combination

_____ 8. The arbor nut is always loosened by turning it in the _____ .
        A. opposite direction the blade rotates
        B. same direction the blade rotates
        C. direction of a right-handed thread
        D. direction of a left-handed thread

_____ 9. A dado set is designed to cut _____ .
        A. dados
        B. grooves
        C. rabbets
        D. all of the above

_____ 10. When circular saw blades become dull they tend to _____ .
        A. burn the material being cut
        B. need more feed pressure of material
        C. have chipped teeth
        D. all of the above

Completion

Complete each sentence by inserting the best answer on the line near the number.

_____ 1. Carbide-tipped blades are so hard that they must be sharpened by _____ -impregnated grinding wheels.

_____ 2. Most of the saw blades being used at present are _____ .

_____ 3. The _____ saw blade is suited for when a variety of cross-cutting and ripping is to be done, and it eliminates the need for changing blades.

_____ 4. A circular saw blade may overheat if the rate of feed is too _____ .

_____ 5. The teeth on a _____ saw blade act like a series of small chisels.

_____ 6. The _____ of a circular saw is designed to keep the tooth that follows it from biting too deeply into the material.

_____ 7. The teeth on a _____ carbide-tipped blade are similar to the rip teeth on a high speed steel blade.

_____ 8. The versatile carbide-tipped _____ blade is probably the most widely used by carpenters.

_____ 9. The valleys between the teeth on a saw blade are known as _____ .

_____ 10. Arbors are always threaded in a direction that will prevent the nut from becoming _____ during operation.

Identification: Blades

Identify each term, and write the letter of the best answer on the line next to each number.

_____ 1. anti-kick back tooth

_____ 2. hook angle

_____ 3. gullet

_____ 4. anti-vibration cuts

_____ 5. square grind

_____ 6. alternate bevel

_____ 7. triple chip

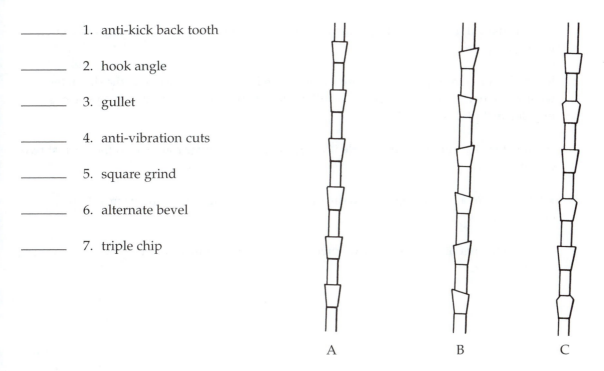

A   B   C

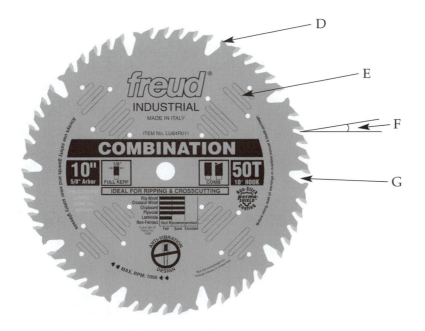

Math Problem-Solving

Solve the following math problems.

_____ 1. What is the value, rounded to the nearest four decimal places, of π in the area of a circle formula $A = \pi r^2$?

_____ 2. If the circumference (perimeter) formula for a circle is $C = \pi d$, where d is the diameter, what is the perimeter of a circle with a diameter of 10? Answer rounded to the nearest two decimal places.

_____ 3. What is the circumference of a circle if the radius is 7? Answer rounded to the nearest two decimal places.

_____ 4. What is the area of a circle if the radius is 12? Answer rounded to the nearest two decimal places.

_____ 5. What is the area of a circle if the diameter is 26? Answer rounded to the nearest two decimal places.

18 Radial Arm and Miter Saw

Multiple Choice

Write the letter for the best answer on the line next to the number of the sentence.

_____ 1. The size of the radial arm saw is determined by the _____ .
   A. the width of the board it can rip
   B. the length of the table
   C. the diameter of the largest blade it can use
   D. the horsepower of the motor

_____ 2. The depth of cut on a radial arm saw is controlled by _____ .
   A. raising or lowering the table
   B. horizontally swinging the arm
   C. tilting the motor unit
   D. raising or lowering the arm

_____ 3. Adjust the depth of cut on the radial arm saw so that the saw blade is about _____ below the surface of the table.
   A. ½″
   B. ⅜″
   C. ¾″
   D. 1/16″

_____ 4. When cutting thick material on the radial arm saw, _____ .
   A. pull the saw through the stock rapidly
   B. pull the saw slowly or hold it back (hesitate) somewhat
   C. never use a stop block
   D. remove the saw's guard

_____ 5. A stop block is fastened to the table of the radial arm saw when _____ .
   A. ripping stock
   B. cutting many pieces the same length
   C. cutting various sized pieces
   D. the saw is in the out-rip position

_____ 6. Upon completing a cut on the radial arm saw, always _____ .
   A. raise the blade up out of the table
   B. return the saw to the starting point behind the fence
   C. place the saw directly in front of the fence
   D. lower the blade to where it contacts the table

_____ 7. The power saw that is most like a radial arm saw is called the _____ .
   A. power miter box
   B. chop box
   C. sliding miter saw
   D. all of the above

_____ 8. The more common sizes of power miter saws are _____ .
- A. 6½″, 7¼″, and 8¼″
- B. 6″, 8″, and 10″
- C. 10″ and 12″
- D. 5″ and 7″

Completion

Complete each sentence by inserting the best answer on the line near the number.

_____ 1. The _____ is typically designed to cut at right angles only.

_____ 2. The _____ saw is specifically designed to crosscut interior and exterior trim.

_____ 3. The arm of a radial arm saw moves horizontally in a complete _____ .

_____ 4. An abrasive or special _____ blade may be used to cut steel studs.

_____ 5. The power miter saw blade may be adjusted to _____, right or left.

_____ 6. The _____ saw allows the blade to chop, tilt, and move on rails as it cuts.

_____ 7. _____ to the saw to determine the best rate of feed for the cut.

_____ 8. When using a radial arm or power miter saw always keep material firmly placed against _____.

Math Problem-Solving

Solve the following math problems.

_____ 1. What is the volume of a cube with a side measuring 16″?

_____ 2. What is the volume of the rectangular solid with a base measuring 4′ × 8′ and a height of 12′?

_____ 3. What is the total area of all sides of the rectangular solid in question 2?

_____ 4. What is the volume of the triangular solid shown if the base area is 15 and the height is 7?

_____ 5. What is the volume of a rectangular solid measuring 6′-4″ wide, 7′-5″ long, and 4′-6″ tall? Answer in terms of feet to the nearest three places.

19 Table Saws

Multiple Choice

Write the letter for the best answer on the line next to the number of the sentence.

_____ 1. _____ when operating the table saw.
   A. Never cut freehand
   B. Always use the rip fence for ripping
   C. Never reach over a running blade
   D. all of the above

_____ 2. A _____ can be cut by making multiple passes across a table saw.
   A. dado
   B. rabbet
   C. V-groove
   D. all of the above

_____ 3. When the miter gauge is turned and the blade is tilted, the resulting cut is a _____ .
   A. flat miter
   B. compound miter
   C. end miter
   D. bevel

_____ 4. _____ are useful aids to hold work against the fence and down on the
              table surface during ripping operations.
   A. Stop blocks
   B. Feather boards
   C. Taper-ripping jigs
   D. Miter jigs

_____ 5. The most common size table saw used on construction sites is the _____ model.
   A. 6″
   B. 8″
   C. 10″
   D. 12″

_____ 6. The table saw is favored over the radial arm saw for _____ .
   A. ripping
   B. crosscutting
   C. miter cuts on long stock
   D. dado cuts

_____ 7. When ripping on the table saw, always use a push stick if the stock is under _____ .
   A. 2″
   B. 3″
   C. 4″
   D. 5″

_____ 8. An auxiliary tabletop can _____ .
   A. prevent thin stock from slipping under the fence
   B. prevent narrow rippings from slipping between the saw blade and the table insert
   C. help prevent accidents
   D. all of the above

Completion

Complete each sentence by inserting the best answer on the line near the number.

_____ 1. The size of the table saw is determined by the diameter of the _____ .

_____ 2. The table saw's blade can be tilted up to a _____ -degree angle.

_____ 3. Handwheels on a table saw are used to adjust the _____ .

_____ 4. During the ripping operation, the stock is guided by the _____ .

_____ 5. The _____ slides in grooves on the table saw's surface.

_____ 6. Blade height is to be adjusted to about _____ above the stock.

_____ 7. A table saw operator should never stand directly in back of the _____ .

_____ 8. Cut stock that is left between a running blade and the fence may result in possible _____ that could injure those in its path.

_____ 9. Ripping on a _____ is done the same as straight ripping, except the blade is tilted.

_____ 10. When crosscutting stock, the _____ is used to guide the stock past the blade.

Math Problem-Solving

Solve the following math problems.

_____ 1. What is the area of a circle with the radius of 16?

_____ 2. In geometry, the base of a prism, such as the cylinder shown, is the side with an opposite, parallel side. What is the height of the prism shown? 50

_____ 3. Using the formula for the volume of a prism, V = Base Area × Height, what is the volume of the figure shown? 18

_____ 4. What is the volume of 5 cylinders with a height of 10 and a radius of 4?

_____ 5. What is the volume of a cylindrical hole that measures 24′-9″ across and 5′ deep?

Discussion

Write your answer(s) on the lines below.

1. Make a list of general safety rules for the table saw.

_____

_____

_____

_____

_____

2. What advantages does the table saw have over the radial arm saw?

_____

_____

_____

_____

_____

_____

20 Understanding Architectural Plans

Multiple Choice

Write the letter for the best answer on the line next to the number of the sentence.

_____ 1. Blueprinting is a process of making copies of construction drawings where the end result is _____ .
   A. called plotting
   B. white lines on a blue background
   C. blue lines on a white background
   D. similar to CAD

_____ 2. CAD stands for _____ .
   A. Custom Associated Drawings
   B. Computer Aided Drafting
   C. Computer Associated Drawings
   D. Created Architectural Drawings

_____ 3. An isometric view is a(n) _____ .
   A. horizontal picture
   B. vertical picture
   C. obtuse view
   D. 3D perspective

_____ 4. When the plan view, elevation, section view, and details are put together they constitute a _____ .
   A. 3D perspective
   B. structure
   C. typical plan
   D. set of construction prints

_____ 5. The perspective of a floor plan is that of a horizontal cut made through the _____ .
   A. foundation, several inches below the sill
   B. floor joists, mid span
   C. floor, just below the finished subfloor
   D. walls, 4 – 5 feet above the floor

_____ 6. Information located in the schedule of a set of prints is _____ .
   A. rough openings of windows
   B. size of doors
   C. type of windows and doors
   D. all the above

_____ 7. Changes to a set of prints must be approved by the _____ .
   A. architect
   B. contractor
   C. foreman
   D. union

_____ 8. The most commonly used scale on a set of prints used to show details of the building site is _____ .
  A. ⅟₁₆″ equals 1 foot
  B. ¼″ equals 1 foot
  C. ½″ equals 1 foot
  D. 1″ equals 1 foot

_____ 9. Line contrast on a drawing is refers to the _____ .
  A. variations in dotted lines
  B. variations in line width
  C. style of the line end points
  D. straightness of the line

_____ 10. A small line that shows where a dimension line begins and ends is called a(n) _____ .
  A. break line
  B. dimension line
  C. extension line
  D. leader line

Completion

Complete each sentence by inserting the best answer on the line near the number.

_____ 1. Multi-view drawings, also called _____ drawings, convey most of the information needed for construction.

_____ 2. The lines in a/an _____ drawing diminish in size as they approach a vanishing point.

_____ 3. Presentation drawings are usually the _____ type.

_____ 4. The _____ plan simulates a view looking down from a considerable height.

_____ 5. The direction and spacing of floor and roof framing is shown on the _____ plan.

_____ 6. The drawings that show the shape and finishes of all sides of the exterior of a building are called _____ .

_____ 7. Window _____ give information about the location, size, and type of windows to be installed in a building.

_____ 8. For complex commercial projects, a _____ guide has been developed by the Construction Specifications Institute.

9. The triangular_____ scale is used to scale lines when making drawings.

_____

10. The most commonly used scale on blueprints is _____ inch(es) equals one foot.

_____

Matching

Write the letter for the best answer on the line near the number to which it corresponds.

_____ 1. pictorial drawings

_____ 2. multi-view

_____ 3. isometric drawing

_____ 4. presentation drawing

_____ 5. plot plan

_____ 6. interior elevations

_____ 7. modular measure

_____ 8. section

_____ 9. specifications

_____ 10. dimensions

A. take precedence over the drawing if a conflict arises

B. used to show the appearance of the completed building

C. a dash is always placed between the foot and the inch

D. most common ones show the kitchen and bath cabinets

E. the horizontal lines are drawn at 30° angles

F. two-dimensional drawings that convey the most information

G. shows information about the lot

H. three-dimensional isometric or perspective drawing

I. buildings designed using a grid with a unit of 4″

J. shows a vertical cut through all or part of a construction

Identification: Lines

Identify each term, and write the letter of the best answer on the line next to each number.

_____ 1. object line

_____ 2. hidden line

_____ 3. centerline

_____ 4. section reference

_____ 5. break line

_____ 6. dimension line

_____ 7. extension line

_____ 8. leader line

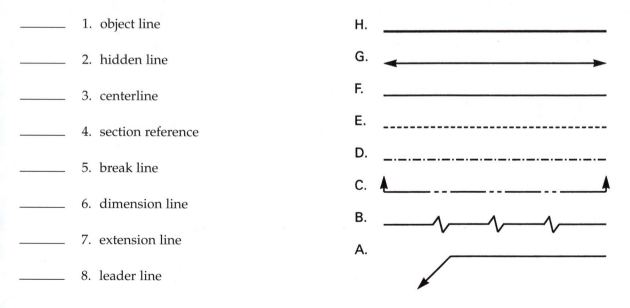

Math Problem-Solving

Solve the following math problems.

_____ 1. What is the measurement to the nearest sixteenth of an inch for 3.33333"?

_____ 2. What is the feet-inches measurement for 3.33333333'?

_____ 3. How many feet are represented by the scaled measurement of 3" when the scale is used is ¼"=1'?

_____ 4. What scaled measurement would be needed to represent 6'-0" in a ¾" = 1' scale?

_____ 5. What total actual measurement is represented by the sum of two scaled measurements of 3³⁄₁₆" and 2⁵⁄₁₆" using a scale of ⅛" = 1'?

21 Floor Plans

Multiple Choice

Write the letter for the best answer on the line next to the number of the sentence.

_____ 1. The least often drawing used by a carpenter is _____ .
   A. electrical plan
   B. truss layout
   C. front elevation
   D. floor plan

_____ 2. A closet door type that only allows one-half the full width of the opening to be open is a _____ .
   A. bypass door
   B. sliding door
   C. bifold door
   D. pocket door

_____ 3. A window with hinges along the top edge is called a _____ .
   A. hopper window
   B. awning window
   C. double hung window
   D. casement window

_____ 4. A scuttle is a(n) _____ .
   A. ceiling opening
   B. attic access
   C. crawl opening
   D. all of the above

_____ 5. The _____ is often found in a window schedule.
   A. window manufacturer
   B. window size
   C. window type
   D. all of the above

_____ 6. Dashed lines on a floor plan indicate a(n) _____ .
   A. location of a hidden girder
   B. interior wall opening without a door
   C. wall cabinet above a base cabinet
   D. all of the above

_____ 7. The direction of rise of stairs is indicated on a set of prints by a(n) _____ .
   A. arrow
   B. left to right rule
   C. left side is usually the top of the stairs
   D. all of the above

_____ 8. The symbol on a set of prints that is a circle with two parallel lines through it indicates a _____ .
   A. smoke detector
   B. receptacle
   C. switch
   D. ceiling light

_____ 9. When dimensions for the location of a door are not given it is assumed the door is _____ .
   A. centered between two walls
   B. centered for the room being entered such as a closet
   C. framed in the corner of the room
   D. all of the above

_____ 10. Dimension lines are draw from the _____ .
   A. outside face of the frame
   B. inside face of concrete
   C. outside face of brick veneer
   D. all of the above

Completion

Refer to the print drawing on page 74 to complete the following sentences.

_____ 1. The overall length of the building is _____ .

_____ 2. The dimensions of the utility room are _____ by _____ .

_____ 3. The fireplace is located in the _____ room.

_____ 4. The exterior door symbol in the dining room is that of a _____ .

_____ 5. The size of the floor joists in the living room and the kitchen is _____ .

_____ 6. In heated areas, all exterior studs are to be _____ placed 16″ on center.

_____ 7. The dimensions of the garage are _____ by _____ .

_____ 8. Not counting the garage door, there are _____ entrance doors into the garage.

_____ 9. The _____ can be entered only through the garage.

_____ 10. The plan is drawn to a scale of _____ = 1′-0″.

_____ 11. A _____ door is on the pantry.

_____ 12. The dimensions of the living room are _____ by _____ .

_____ 13. In the living room, the number of electrical outlets that can be controlled by switches is _____ .

_____ 14. Attic access is located in the _____ .

Sketch: Floor Plan Symbols

Sketch each described item in the space provided.

1. exterior sliding door

6. awning window

2. pocket door

7. sliding window

3. bifold door

8. water closet

4. double-hung window

9. standard tub

5. casement window

10. standard shower

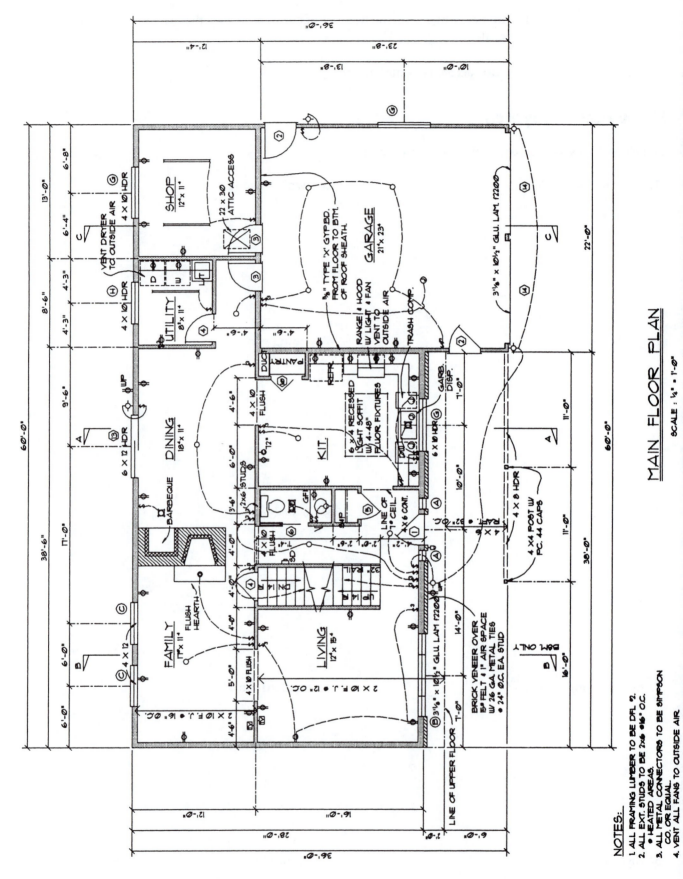

MAIN FLOOR PLAN

SCALE : ¼" = 1'-0"

NOTES:
1. ALL FRAMING LUMBER TO BE DFL #2.
2. ALL EXT. STUDS TO BE 2x6 @16" O.C.
   • HEATED AREAS.
3. ALL METAL CONNECTORS TO BE SIMPSON
   CO. OR EQUAL.
4. VENT ALL FANS TO OUTSIDE AIR

GARAGE
21' x 23'

SHOP
12' x 11'

UTILITY
8' x 11'

DINING
18' x 11'

KIT

PANTRY

FAMILY
17' x 11'

LIVING
12' x 15'

BARBEQUE

FLUSH HEARTH

22 x 30 ATTIC ACCESS

⅝" TYPE "X" GYP.BD.
FROM FLOOR TO BTM.
OF ROOF SHEATH.

VENT DRYER
TO OUTSIDE AIR

RANGE & HOOD
W/ LIGHT & FAN
VENT TO
OUTSIDE AIR

TRASH COMP.

GARB. DISP.

6 x4 RECESSED
LIGHT SOFFIT
W/ 4-48"
FLUOR FIXTURES

LINE OF
2" CEIL.

2 x6 STUDS

4 X 10 FLUSH

4 X 10 FLUSH

4 x 12

4 X 10 FLUSH

4 X 10 FLUSH

4 x 12 HDR

6 x 12 HDR

4 X 10 HDR

4 X 10 HDR

6 X 10 HDR

4 X 8 HDR

4 X 4 POST W/
PC. 44 CAPS

3⅛" x 10½" GLU. LAM. F2200

3⅛" x 10½" GLU. LAM. F2200

3⅛" x 10½" GLU. LAM. F2200

BRICK VENEER OVER
15# FELT & 1" AIR SPACE
W/ 26 GA. METAL TIES
@ 24" O.C. EA STUD

2 x 10 F.J. @ 16" O.C.

2 x 10 F.J. @ 16" O.C.

2 x 6 F.J. @ 12" O.C.

DBL ONLY

LINE OF UPPER FLOOR

36'-0"
12'-0"
23'-8"
13'-8"
10'-0"
60'-0"
6'-8"
13'-0"
6'-4"
8'-6"
4'-3"
4'-3"
9'-6"
38'-6"
11'-0"
6'-0"
6'-0"
22'-0"
60'-0"
10'-0"
11'-0"
11'-0"
36'-0"
14'-0"
16'-0"
7'-0"
12'-0"
16'-0"
28'-0"
7'-0"
6'-0"
36'-0"

74

Math Problem-Solving

Provide the best answer for each of the following questions.

1. The length of line A is _____ inches.

2. The length of line B is _____ inches.

3. The length of line C is _____ inches.

4. The length of line D is _____ inches.

5. The length of line E is _____ inches.

6. The length of line F is _____ inches.

7. The length of line G is _____ inches.

8. The length of line H is _____ inches.

9. The length of line I is _____ inches.

10. The length of line J is _____ inches.

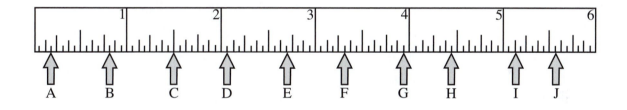

Discussion

Write your answer(s) on the lines below.

1. Refer to the figure on page 74 again. If you were building the home on the floor plan for yourself to live in, what changes might you make to improve the design of the home?

_____

_____

_____

22 Sections and Elevations

Multiple Choice

Write the letter for the best answer on the line next to the number of the sentence.

_____ 1. A vertical cut view through a building is referred to as a _____ .
- A. plan view
- B. section view
- C. detail view
- D. elevation view

_____ 2. The information least likely to be found in a section view is _____ .
- A. ceiling insulation
- B. wall sheathing
- C. a dimension between windows
- D. ceiling height

_____ 3. Elevations of a set of prints show the view of a building from _____ .
- A. above
- B. below
- C. outside
- D. inside looking out

_____ 4. On what drawing would you most likely find the fewest dimensions to locate a chimney?
- A. Floor plan
- B. Elevation
- C. Foundation plan
- D. Detail

_____ 5. Dotted lines drawn on the glass of a window indicates the window is a _____ .
- A. fixed window
- B. slider
- C. hung window
- D. hinged window

Completion

Base your answers to questions 1–10 on the figure on page 79.

_____ 1. While floor plans are views of a horizontal cut, sections show _____ cuts.

_____ 2. Sections are usually drawn at a scale of _____ = 1'-0".

_____ 3. _____ are cut across the width or through the length of the entire building.

_____ 4. Enlargements of part of a section, which are done to convey additional needed information, are called _____ .

_____ 5. According to section A-A, the bottom of the footer must be located _____ inches below grade.

_____ 6. The insulation between the crawl space and the subfloor has an R value of _____ in section A-A.

_____ 7. In section A-A, a minimum distance of _____ inches must be maintained between the .006 black vapor barrier and the bottom of the floor joists.

_____ 8. The insulation in the attic has a thickness of _____ inches in section A-A.

_____ 9. The roof has a _____-12 slope in section A-A.

_____ 10. The plywood sheathing on the roof calls for a thickness of _____ in section A-A.

Base your answers to questions 11–17 on the figure on page 80.

_____ 11. The rafters are framed with _____ by _____ stock in section A-8.

_____ 12. The thickness of the basement floor is _____ in section A-8.

_____ 13. In section A-8, the basement wall is _____ inches thick.

_____ 14. Section A-8 is drawn to a scale of _____ = 1′-0″.

_____ 15. The insulation on the outside of the basement wall is _____ inches thick in section A-8.

_____ 16. In section A-8, the floor joists are framed with _____ by _____ stock.

_____ 17. The footer in section A-8 is _____ inch(es) thick.

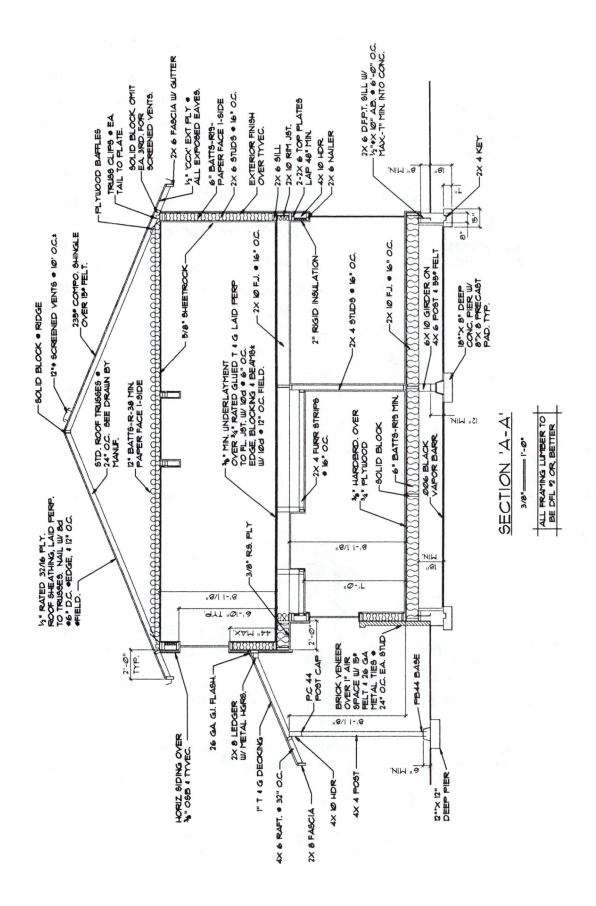

SECTION 'A-A'

3/8" ━━━ 1'-0"

ALL FRAMING LUMBER TO
BE DFL #2 OR BETTER

Labels (reading around the drawing):

- SOLID BLOCK @ RIDGE
- PLYWOOD BAFFLES
- 12"⌀ SCREENED VENTS @ 10' O.C.±
- TRUSS CLIPS @ EA.
- SOLID BLOCK OMIT EA 3RD. FOR SCREENED VENTS.
- 2X 6 FASCIA W/ GUTTER
- 235# COMPO. SHINGLE OVER 15# FELT.
- ½" 'CCX' EXT PLY @ ALL EXPOSED EAVES.
- 6" BATTS-R19- PAPER FACE I-SIDE
- 2X 6 STUDS @ 16" O.C.
- EXTERIOR FINISH OVER TYVEC.
- 2X 6 SILL
- 2X 10 RIM JST.
- 2-2X 6 TOP PLATES LAP 48" MIN.
- 4X 10 HDR
- 2X 6 NAILER
- 2X 6 D.F.P.T. SILL W/ ½"⌀X 10" A.B. @ 6'-0" O.C. MAX.-7" MIN. INTO CONC.
- 2X 4 KEY
- 5TD. ROOF TRUSSES @ 24" O.C. SEE DRAWN BY MANIF.
- 12" BATTS-R-38 MIN. PAPER FACE I-SIDE
- 5/8" SHEETROCK
- ⅝" MIN. UNDERLAYMENT OVER ¾" RATED GLUED T & G LAID PERP TO FL. JST. W/ 10d @ 6" O.C. EDGE, BLOCKING & BEAMS & W/ 10d @ 12" O.C. FIELD.
- 2X 10 F.J. @ 16" O.C.
- 2" RIGID INSULATION
- 2X 4 STUDS @ 16" O.C.
- 2X 10 F.J. @ 16" O.C.
- 6X 10 GIRDER ON 4X 6 POST & 55# FELT
- 18"X 8" DEEP CONC. PIER W/ 8"X 8 PRECAST PAD. TYP.
- HORIZ SIDING OVER ⅜" OSB & TYVEC.
- ½" RATED 32/16 PLY. ROOF SHEATHING, LAID PERP. TO TRUSSES. NAIL W/ 8d @6" D.C. @EDGE, & 12" O.C. @FIELD.
- 2X 4 FURR STRIPS @ 16" O.C.
- ⅜" HARDBRD. OVER ¾" PLYWOOD
- SOLID BLOCK
- 6" BATTS-R19 MIN.
- .006 BLACK VAPOR BARR.
- 3/8" R.S. PLY
- 26 GA. G.I. FLASH.
- 2X 8 LEDGER W/ METAL HGRS.
- 1" T & G DECKING
- PC 44 POST CAP
- BRICK VENEER OVER 1" AIR SPACE W/ 15# FELT & 26 GA METAL TIES @ 24" O.C. EA STUD.
- PB44 BASE
- 12"X 12" DEEP PIER
- 4X 6 RAFT. @ 32" O.C.
- 2X 8 FASCIA
- 4X 10 HDR
- 4X 4 POST
- 2'-0" TYP.
- 6'-10" TYP.
- 8'-11⅛"
- 44" MAX.
- 2'-0"
- 8'-11⅛"
- 8'-1⅛"
- 1'-0"
- 9'-11⅛"
- 8" MIN.
- 18"
- 15"
- 8"
- 3 MIN.
- 12 MIN.
- 3 MIN.
- 6" MIN.
- 3 MIN.

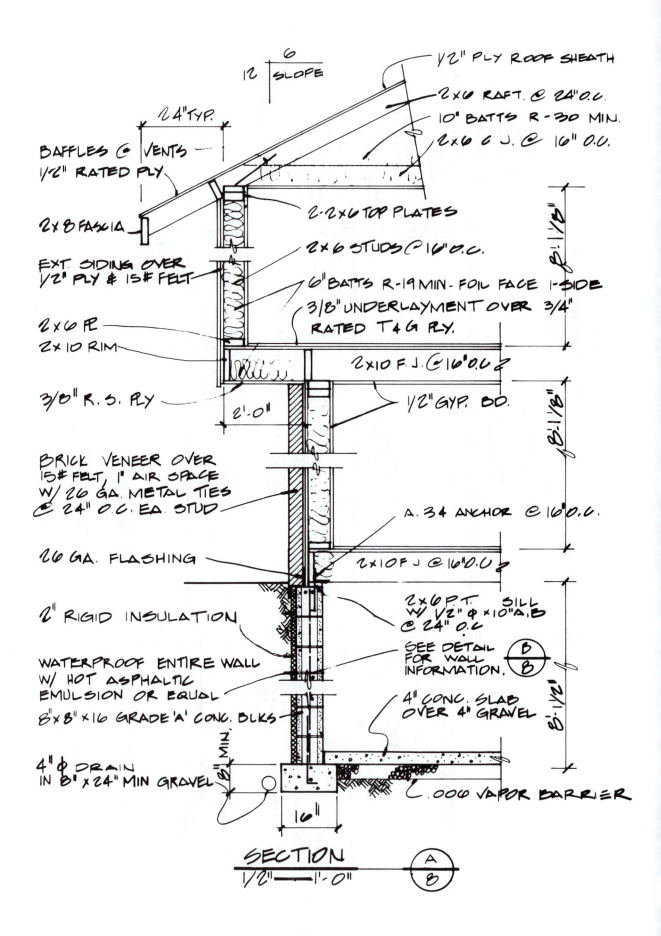

$\frac{6}{SLOPE}$ 12

1/2" PLY ROOF SHEATH

2×6 RAFT. @ 24" O.C.

10" BATTS R-30 MIN.

2×6 C.J. @ 16" O.C.

BAFFLES @ VENTS —
1/2" RATED PLY

24" TYP.

2×8 FASCIA

2·2×6 TOP PLATES

2×6 STUDS @ 16" O.C.

6" BATTS R-19 MIN. FOIL FACE I-SIDE
3/8" UNDERLAYMENT OVER 3/4"
RATED T&G PLY.

EXT. SIDING OVER
1/2" PLY & 15# FELT

2×6 PL.
2×10 RIM

2×10 F.J. @ 16" O.C.

3/8" R.S. PLY

2'-0"

1/2" GYP. BD.

8'-11/8"

8'-11/8"

BRICK VENEER OVER
15# FELT, 1" AIR SPACE
W/ 26 GA. METAL TIES
@ 24" O.C. EA. STUD

A. 34 ANCHOR @ 16" O.C.

26 GA. FLASHING

2×10 F. J @ 16" O.C.

2" RIGID INSULATION

2×6 P.T. SILL
W/ 1/2" Ø × 10" A.B.
@ 24" O.C.

SEE DETAIL
FOR WALL
INFORMATION.   B / B

8'-1/2"

WATERPROOF ENTIRE WALL
W/ HOT ASPHALTIC
EMULSION OR EQUAL

4" CONC. SLAB
OVER 4" GRAVEL

8"×8"×16 GRADE 'A' CONC. BLKS

4" Ø DRAIN
IN 8" × 24" MIN GRAVEL

6" MIN.

.006 VAPOR BARRIER

16"

SECTION
1/2" ▬ 1'-0"   A / 8

Refer to the figures on pages 79–81 to complete the following sentences.

_____ 18. According to the accompanying detail, there are _____ inches of gravel under the concrete floor.

_____ 19. The gypsum wallboard on the detail drawing has a thickness of _____ inch(es).

_____ 20. The thickness of the insulation of the exterior wall in the figure on page 81 is _____ inch(es).

_____ 21. Elevations are usually drawn at the same scale as the _____ .

_____ 22. There are usually _____ elevations in each set of drawings.

_____ 23. In relation to other drawings, elevations have few _____ .

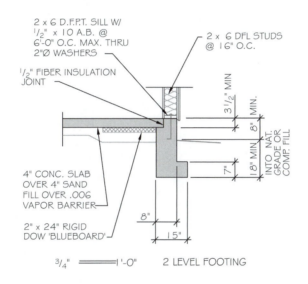

Math Problem-Solving

Solve the following math problems.

_____ 1. What is the sum of $7^3 + 14$?

_____ 2. What is the value of $16^2 \times 16^3$?

_____ 3. What is the value of $6^2 + 8^2$?

_____ 4. What is the square root of 100?

_____ 5. Find the square root of $(14^2 + 15^2)$. Answer rounded to two decimal places.

23 Plot and Fountation Plans

Multiple Choice

Write the letter for the best answer on the line next to the number of the sentence.

_____ 1. The plot plan must show _____ .
   A. a frontal view of the completed structure
   B. the direction and spacing of the framing members
   C. compliance with zoning and health regulations
   D. all of the above

_____ 2. Metes and bounds on a plot plan refer to _____ .
   A. elevation
   B. boundary lines
   C. utility easements
   D. the slope of the finish grade

_____ 3. When contour lines are spaced close together, _____ .
   A. a new grade is being indicated
   B. a gradual slope is indicated
   C. they are closer to sea level
   D. a steep slope is indicated

_____ 4. Instead of contour lines, the slope of the finish grade can be indicated by _____ .
   A. an arrow
   B. metes and bounds
   C. easements
   D. the measurement of rods

_____ 5. The distance from the property line to the building is known as _____ .
   A. the point of beginning
   B. setbacks
   C. variances
   D. bearings

_____ 6. The location of the footer on a foundation plan is indicated by _____ .
   A. solid lines
   B. solid lines with a dash in the center
   C. dashed lines
   D. object lines

_____ 7. A recess in a foundation wall to support a girder is called a _____ .
   A. benchmark
   B. pocket
   C. casement
   D. seat

_____ 8. On foundation plans, the walls are dimensioned from the _____ .
   A. centerlines of opposing walls
   B. inside of one wall to the inside of the next
   C. outside area of the footer
   D. face of one wall to the face of the next

Completion

Refer to the following figure to complete the sentences.

_____ 1. The elevation of the finished floor on the accompanying plot plan is _____ .

_____ 2. The setback distance of the home from the front boundary line is _____ on the plot plan.

_____ 3. The plot plan is drawn on a scale of 1″ = _____ .

_____ 4. The dimensions of the house are roughly _____ by _____ .

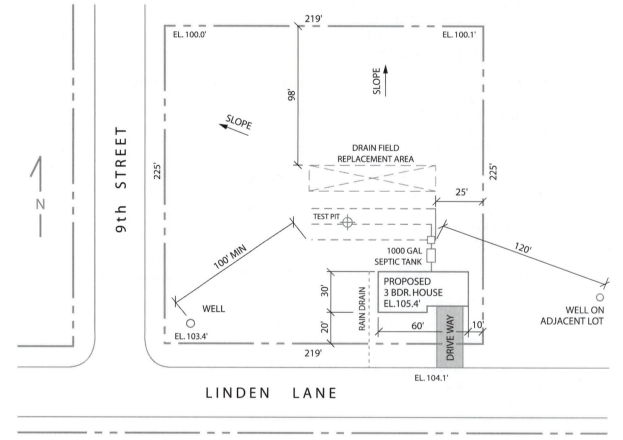

_____ 5. The longest property boundary on the plot plan is _____ feet.

_____ 6. The driveway connects to road called _____ .

_____ 7. The compass direction of the well from the house is _____ .

_____ 8. The compass direction that the rain drain allows water to flow is most likely _____ .

_____ 9. The direction the sewer system is from the house is _____ .

_____ 10. The west property line is _____ feet long.

Math Problem-Solving

Solve the following math problems.

_____ 1. Convert 97.40′ to feet-inches to the nearest sixteenth of an inch.

_____ 2. Convert 34.56′ to feet-inches to the nearest eighth of an inch.

_____ 3. What is the elevation of the foundation if the Finish Floor Height is 98.50′ and the floor system thickness is 1′–1⅜″? Answer in terms of feet to the nearest two decimal places.

_____ 4. How many degrees is N 45°E from South?

_____ 5. Add 23° 30′ + 34° 50′ + 45° 45′. Answer in degrees-minutes.

Discussion

Write your answer(s) on the lines below.

1. What are some of the reasons municipal planning officials insist on a plot plan prior to issuance of a building permit?

_____

_____

_____

_____

24 Building Codes and Zoning Regulations

Multiple Choice

Write the letter for the best answer on the line next to the number of the sentence.

_____ 1. Similar size and purpose buildings are limited to various areas of cities and towns by _____ .
   A. building codes
   B. construction techniques
   C. zoning regulations
   D. labor unions

_____ 2. Green space refers to the _____ .
   A. minimum lot width
   B. amount of landscaped area
   C. structure's maximum ground coverage
   D. off-street parking required

_____ 3. Structures built prior to zoning regulations that do not exist within their proper zone are called _____ .
   A. unreformed
   B. nonsanctioned
   C. nonconfirming
   D. preapproved

_____ 4. Hardships imposed by zoning regulations may be relieved by a _____ granted by the zoning board.
   A. dissention
   B. objection
   C. variance
   D. reassessment

_____ 5. Minimum standards of safety concerning the design and construction of buildings are regulated by _____ .
   A. zoning laws
   B. an appeals committee
   C. construction costs
   D. building codes

_____ 6. The national code created from a blend of three other national codes is called _____ .
   A. International Residential Code
   B. Uniform Building Code
   C. Standard Building Code
   D. National Building Code

_____ 7. An area of importance in residential building codes is _____ .
   A. exit facilities
   B. room dimensions
   C. requirements for bath, kitchens, and hot and cold water
   D. all of the above

_____ 8. The building permit fee is usually based on the _____ .
      A. square footage of the building
      B. estimated cost of construction
      C. occupant load
      D. location of the building

_____ 9. Foundation inspections occur prior to the _____ .
      A. placement of concrete
      B. erection of forms
      C. placement of reinforcement rod
      D. removal of the forms

_____ 10. It is the responsibility of the _____ to notify the Building Official when it is time for a scheduled inspection.
      A. loan officer
      B. zoning officer
      C. contractor
      D. business agent

Math Problem-Solving

Solve the following math problems.

_____ 1. What is always the sum of the three angles inside any triangle?

_____ 2. If a right triangle has a second inside angle of 35°, what is the measure of the third angle?

_____ 3. What is the third angle inside a triangle with two angles measuring 49° 30′ and 20° 10′? Answer in terms of degree-minutes.

_____ 4. What is the angle formed between two perpendicular intersecting lines?

_____ 5. If an equilateral triangle has equal interior angles, what is the measure of one of the angles?

Discussion

Write your answer(s) on the lines below.

1. Although zoning is intended to protect the rights of the property owners, how might it at the same time infringe on these very rights?

   _____

   _____

   _____

2. For the most part, a good rapport exists between inspectors and builders. Why is it important for those entering the trade to be aware of this and strive to continue it?

   _____

   _____

   _____

Section 1: Building for Success

IMPLEMENTING A SAFETY PROGRAM

At the job site or in the school construction lab, no objective should neglect to take into consideration the safety and welfare of workers and students. Completion dates, schedules, quality, and profits are important goals in the construction industry. However, none should be attained at the expense of people's lives. The highest priority on the job has to be an effective plan to ensure worker safety. The safety program has to be ongoing from day one to be effective. The responsibility of this program is everyone's. Both skilled and unskilled men and women who subject themselves to numerous hazards and potential accidents while working deserve the best protection available.

In most cases the contractor has informed his or her employees of appropriate job safety. Acceptable safety practices are reinforced by insurance companies and local, state, and national agencies such as the Occupational Safety and Health Administration (OSHA).

A company safety program must be well researched, prepared, and presented to the employees so they see it is for their benefit. Implementation must become part of everyone's daily tasks. The basic premise should be to greatly reduce the probability of construction-related accidents. Safety research should be conducted using all available resources, including statistics depicting the types and causes of construction accidents. The collected information should then be used to structure and implement the safety program.

Much research information may be obtained from insurance companies, labor statistics, OSHA, and the National Safety Council. A desired result of the research is to prepare a safety program that the employees will accept. The focus will center on the workers' welfare. Realistically this concentration should be a high priority with all workers.

Employees have to be involved from the beginning. The success of the program depends on how much the workers themselves believe in staying healthy. It has to be a program that automatically kicks in on a daily basis. A safe working environment has to be important to every worker or student on the job. This calls for influence on the part of instructors or contractors. They will have to "sell" the safety plan to employees. Almost everyone has listened to radio station "WIIFM" (What's in it for me?). That will be the key for acceptance. This connection has to be made if the workers are to accept the plan as a part of their job and security.

The annual number of injuries and fatalities in the construction industry share the lead with those in mining and farming. The very nature of these professions exposes people to a high risk for accidents. Each trade represented in construction has specific high-risk tasks that endanger workers. The overall safety program should become an element in all construction jobs.

Discussions, videos, speakers, demonstrations, and literature can be a vital part of an effective safety program. Each student or worker can be given a part in the presentation, application, or feedback report of a company safety program. As potential accidents are discovered and studied by workers, attention can be drawn to accident prevention.

Hazards and dangerous working conditions can be eliminated from the work site through employee involvement. Potential subjects for good safety programs are numerous due to the multitude of construction related jobs. Concentrations can be on excavating, shoring, hauling, lifting, cutting, climbing, driving, assembling, electrical grounding, or myriad other activities. Ideally the studies should center on tasks that are carried out by the people on a daily basis, then go on to other more infrequent tasks.

OSHA guidelines are usually used to create a safety program that informs workers and keeps a safety program ongoing. The program should have the safety of the workers in mind. Many construction companies take pride in the length of time between accidents on the job. An incentive program helps everyone become strongly committed to worker safety.

A good safety program will answer the important questions of Who, Why, What, When, Where, and How (in no particular order). To answer these questions: (1) identify the need and target population (Who); (2) justify the safety program with reason (Why); (3) relate the safety program concept (What); (4) establish a time frame for the program (When); (5) list the various work sites for implementation (Where); and (6) clarify the implementation process for everyone (How). Then be prepared to document, report, and reinforce the use of the safety program. A successful safety program will always involve the target population in the planning process. Ownership at this point is extremely important if goals are to be reached. The business owners and/or managers may see how a good safety record can ensure staying on schedule or assure profits for the company, but the employees may view it differently. The workers will be interested in assisting the company with their goals of quality and profits. They will become more highly motivated, however, by a safety program that centers on their health and welfare. That's when the buy-in takes place.

As employers reflect on the time and cost of an effective safety program, they will see and feel the benefits in satisfied workers, lower insurance rates, fewer accidents, less downtime, and healthier work climates. The employees in turn will feel they are a valued part of "their" company.

Focus Questions:
For individual or group discussion

1. What might be the reasons for implementing a safety program at school or on the job?
2. In order for a safety program to be effective, people must see how it can benefit them. How could you as a contractor persuade them to buy in to this type of program?
3. As you introduce a safety program, how might you answer the questions Who, Why, What, When, Where, and How?

SECTION TWO

ROUGH CARPENTRY

25 Leveling and Layout Tools

Multiple Choice

Write the letter for the best answer on the line next to the number of the sentence.

_____ 1. The leveling tool that uses a liquid to determine level is called a _____ .
   A. water level
   B. spirit level
   C. builders level
   D. all of the above

_____ 2. The optical level best suited to determine differences in elevations, when used with a leveling rod, is a(n) _____ .
   A. builders level
   B. transit level
   C. automatic level
   D. all of the above

_____ 3. The difference between a builder's level and a transit is that the builders level _____ .
   A. measures horizontal angles only
   B. has a telescope that tilts up and down
   C. more accurate than the transit
   D. all of the above

_____ 4. To adjust a transit to level that has a 4 leveling screw base, the opposite screws are _____ .
   A. rotated in opposite directions
   B. rotated in the same direction
   C. tightly turned
   D. kept loosely turned

_____ 5. The horizontal line of the crosshairs of a builder's level is used to sight _____ .
   A. elevation differences
   B. horizontal angles
   C. vertical angles
   D. all of the above

_____ 6. The vertical line of the crosshairs of a builder's level is used to sight _____ .
   A. elevation differences
   B. horizontal angles
   C. vertical angles
   D. all of the above

_____ 7. Point A measurement of a builder's level and rod is 42″. Point B measures 62″. Point A is _____ .
   A. 20″ lower than point B
   B. 20″ higher than point B
   C. 104″ higher than point B
   D. 104″ lower than point B

_____ 8. Point A measurement of a builder's level and rod is 42″. Point B measures 62″ while the rod is upside down. Point A is _____ .
   A. 20″ lower than point B
   B. 20″ higher than point B
   C. 104″ higher than point B
   D. 104″ lower than point B

Completion

Complete each sentence by inserting the best answer on the line near the number.

_____ 1. If no other tools are available, a long straightedge and a _____ may be used to level across the building area.

_____ 2. An accurate tool dating back centuries and used for leveling from one point to another is the _____ level.

_____ 3. The _____ level consists of a telescope mounted in a fixed horizontal position with a spirit level attached.

_____ 4. Automatic levels have an internal compensator that uses _____ to maintain a true level line of sight.

_____ 5. Before a level can be used, it must be placed on a _____ or some other means of support.

_____ 6. When adjusting any level, never apply excessive pressure to the _____ .

_____ 7. Because the tape measure is flexible, it is often backed with a _____ to make it more rigid when it is used as a target.

_____ 8. The _____ is the ideal target for longer sightings because of its clearer graduations.

_____ 9. A starting point of known elevations used to determine other elevations is known as a _____ .

_____ 10. When the base of the rod is at the desired elevation, the reading on the rod is known as _____ rod.

_____ 11. The height of an instrument is determined by placing the rod on the _____ and adding that reading to the elevation of the benchmark.

_____ 12. When recording readings of elevation differences, all _____ sights are known as plus sights.

_____ 13. When the level must be set up directly over a particular point, a _____ is attached to a hook centered below the instrument.

_____ 14. A horizontal circle scale on the instrument is divided into quadrants of _____ degrees each.

_____ 15. The horizontal vernier is used to read _____ of a degree.

_____ 16. When laying out a horizontal angle, the instrument must be centered and leveled over the _____ of the angle.

_____ 17. By rotating a full _____ degrees, the laser level creates a level plane of light.

_____ 18. A battery-powered electronic sensor is attached to the leveling rod to detect _____ .

_____ 19. Laser level safety requires that whenever possible the laser be set up so that it is above or below _____ level.

_____ 20. All _____ instruments are required to have warning labels attached to them.

Matching

Write the letter for the best answer on the line near the number to which it corresponds.

_____ 1. water level

_____ 2. builder's level

_____ 3. transit level

_____ 4. horizontal crosshairs

_____ 5. vertical crosshairs

_____ 6. foresights

_____ 7. horizontal circle scale

_____ 8. vertex

_____ 9. laser level

_____ 10. suspended ceiling grids

A. minus sights

B. the point of an angle

C. magnetic or clip on targets are attached to

D. may rotate up to 40 RPS

E. outside ring

F. has a horizontally fixed telescope

G. crosshairs used when laying out angles

H. limited by the length of the plastic tube

I. telescope can be moved up and down

J. crosshairs used for reading elevations

Math Problem-Solving

Solve the following math problems.

_____ 1. If a benchmark elevation is given as 204.5' and a transit reads 30" on a ruler placed on the benchmark, what is the height of the instrument in feet-inches?

_____ 2. If a second reading of another object, in the same setup as in question 1, is 48", the second object is _____ (*higher/lower*) than the bench mark?

Use the table to answer the following questions.

Setup	Backsight	Foresight
1	4'-4½"	5'-5½"
2	6'-9¾"	3'-10"
3	5'-2⅛"	2'-4⅜"
Sum		

_____ 3. What is the sum of the backsights in the chart?

_____ 4. What is the sum of the minus sights?

_____ 5. What is the elevation difference measured from starting point to ending point?

_____ 6. The starting point is _____ (*higher/lower*) than ending point.

26 Laying Out Foundation Lines

Multiple Choice

Write the letter for the best answer on the line next to the number of the sentence.

_____ 1. It is usually the responsibility of the _____ to lay out building lines.
   A. mason
   B. architect
   C. carpenter
   D. surveyor

_____ 2. Before any layout can be made, it is important to determine _____ .
   A. the dimensions of the building and its location on the site from the plot plan
   B. a starting corner by the 6-8-10 method
   C. a level plane to measure from
   D. a proper benchmark

_____ 3. Measure in on each side from the front property line the specified setback to establish the
   _____ .
   A. approximate boundaries of the property
   B. benchmark
   C. the front line of the building
   D. different elevation points

_____ 4. In the absence of a transit or builder's level, a right triangle may be laid out using the
   _____.
   A. speed square
   B. water level
   C. plot plan
   D. Pythagorean Theorem

_____ 5. The diagonal of a rectangular building that measures 34'-6" × 46'-0" is _____ .
   A. 57'-5"
   B. 57'-6"
   C. 57'-6¾"
   D. 80'-6"

_____ 6. If the lengths of opposite sides of a rectangular layout are equal and the diagonal
   measurements are also equal, then the corners are _____ .
   A. parallel
   B. square
   C. divergent
   D. obtuse or acute angles

_____ 7. All corner stakes are located by measuring from _____ .
   A. diagonal points
   B. a benchmark
   C. a point of beginning
   D. the established front and side building lines

_____ 8. _____ are wood frames to which building lines are secured.
   A. Grade rods
   B. Batter boards
   C. Line anchors
   D. Transome boards

_____ 9. Ledgers are usually _____ .
   A. vertical 2″ × 4″ stakes
   B. horizontal 1″ × 6″ boards
   C. precast concrete
   D. fastened below the top of the footer

_____ 10. Batter boards must be erected in such a manner that they _____ .
   A. will last for years
   B. will withstand sideways force when the footer is poured
   C. will not be disturbed during excavation
   D. can support large amounts of downward pressure

Math Problem-Solving

Solve the following math problems.

_____ 1. Determine the solution of $\sqrt{(24^2 + 38^2)}$. Round to nearest two decimal places.

_____ 2. Given the legs of a right triangle are 36 and 48, what is the measure of the longest side?

_____ 3. What is the diagonal of a rectangle with sides that measure 34′-9″ and 42′-2″ in terms of feet-inches to the nearest sixteenth?

_____ 4. Using the 3-4-5 right triangle concept, what is the length of the hypotenuse if the sides are 39 and 52 respectively?

_____ 5. If the diagonal of a square is 9.8995 feet, what is the measurement of the side of the square? Round answer to nearest three decimal places.

Discussion

Write your answer(s) on the lines below.

1. How important is the knowledge of layout instruments and an understanding of plans to the foundation layout?

   _____

   _____

2. List and discuss some of the serious problems that could happen if a mistake were to occur during foundation layout.

   _____

   _____

27 Characteristics of Concrete

Multiple Choice

Write the letter for the best answer on the line next to the number of the sentence.

_____ 1. Concrete form construction is usually the responsibility of the _____ .
   A. laborers
   B. masons
   C. carpenters
   D. ready-mix plant

_____ 2. A chemical reaction called _____ causes cement to harden.
   A. aggregation
   B. aeration
   C. adhesion
   D. hydration

_____ 3. Hardening of concrete can continue for _____ .
   A. days
   B. weeks
   C. months
   D. years

_____ 4. A bag of portland cement contains one cubic foot and weighs _____ pounds.
   A. 70
   B. 82
   C. 94
   D. 100

_____ 5. Type IA is an air-entraining cement used to _____ .
   A. improve resistance to freezing and thawing
   B. withstand great compression
   C. seal oil wells
   D. withstand high temperatures

_____ 6. The quality of concrete is greatly affected by _____ .
   A. the water-cement ratios
   B. rapid evaporation of the water
   C. freezing of the water
   D. all of the above

_____ 7. Aggregates serve as a _____ in concrete.
   A. bonding agent
   B. stabilizer
   C. filler
   D. corrosion inhibitor

_____ 8. A cubic yard contains _____ cubic feet.
   A. 27
   B. 81
   C. 36
   D. 18

_____ 9. Concrete must be delivered within _____ after water has been added to the mix.
   A. 30 minutes
   B. 3½ hours
   C. 15 minutes
   D. 1½ hours

_____ 10. Steel bars are added to concrete to increase its _____ .
   A. compressive strength
   B. tensile strength
   C. resistance to freezing
   D. curing time

Completion

Complete each sentence by inserting the best answer on the line near the number.

_____ 1. A #6 rebar has a diameter of _____ inch(es).

_____ 2. Welded wire mesh is identified by the gauge and spacing of the _____ .

_____ 3. To ease their removal, the inside surfaces of forms are brushed with _____ .

_____ 4. Concrete without admixtures having a slump of greater than _____ inches should not be used.

_____ 5. Vibrating or hand-spading is done to eliminate voids or _____ in the concrete.

_____ 6. Excessive vibration causes concrete to become more liquid, causing more _____ on forms.

_____ 7. Flooding or constant sprinkling of the surface with water after the concrete has set is the most effective method of _____ concrete.

_____ 8. Permanent damage is almost certain if the concrete becomes _____ within the first 24 hours of being placed.

_____ 9. In cold weather, _____ are sometimes added to the concrete to shorten the setting time.

_____ 10. Concrete should be protected from freezing for at least _____ days.

Math Problem-Solving

Solve the following math problems.

_____ 1. What is 3/7 of 21?

_____ 2. If a yard is equal to 3′, how many cubic feet in one cubic yard?

_____ 3. What is the maximum aggregate size allowed in concrete slab that is 6″ thick?

_____ 4. What is the maximum aggregate size allowed when footing rebars are placed 4″ from the edge of the form?

_____ 5. To make a small batch of 4500 psi concrete, how much water should be used with 1½ bags of Portland cement?

Discussion

Write your answer(s) on the lines below.

1. Why is it so important for the carpenter to have a knowledge of concrete?

_____

_____

_____

_____

Name_____   Date _____

28 Forms for Footings, Slabs, Walks, and Driveways

Multiple Choice

Write the letter for the best answer on the line next to the number of the sentence.

_____   1. Slab-on-grade construction permits the structure to have _____ .
    A. lower construction costs
    B. a basement
    C. a crawl space
    D. a higher profile

_____   2. For good drainage with slab-on-grade construction, the top of the slab should be
    _____ .
    A. not more than 4″ below grade
    B. level with the grade
    C. not more than 4″ above grade
    D. not less than 8″ above grade

_____   3. The soil under a slab is sometimes treated with chemicals to _____ .
    A. reduce settling
    B. prevent frost from lifting the slab
    C. control termites
    D. improve drainage

_____   4. Prior to pouring the slab for a slab-on-grade structure, _____ .
    A. a vapor barrier should be installed
    B. all water and sewer lines must be installed
    C. top soil must be removed
    D. all of the above

_____   5. With a monolithic slab, _____ .
    A. no footer is necessary
    B. the slab and footer are one piece
    C. a basement is included
    D. no reinforcement is needed in the slab

_____   6. In an area where the ground freezes to an appreciable depth, the type of slab-on-
    grade construction that is often used is known as _____ .
    A. a monolithic slab
    B. an independent slab
    C. a detached slab
    D. a thickened edge slab

_____   7. To reduce heat loss, rigid insulation is placed _____ .
    A. on top of the slab
    B. under the footing
    C. under the perimeter of the slab
    D. all of the above

_____ 8. Forms for walks and driveways should be built so that _____ .
   A. water will drain from their surfaces
   B. they may be left in place to save on labor cost
   C. the concrete will rest on sod when possible
   D. all of the above

_____ 9. When building forms for curved walks or driveways, it is helpful to _____ .
   A. use hardwood
   B. install the grain horizontally if using plywood
   C. wet the stock before bending
   D. prestress the form for several days prior to the pour

_____ 10. When bending curves with a long radius, _____ may be used for forms.
   A. 2″ × 4″s
   B. ridged insulation
   C. 1″ × 4″s
   D. A and B

Completion

Complete each sentence by inserting the best answer on the line near the number.

_____ 1. The _____ for a foundation provides a base to spread the load of the structure over a wide area of the earth.

_____ 2. To provide support for posts and columns, a _____ footing is used.

_____ 3. In residential construction, the width of the footing is usually _____ the thickness of the wall.

_____ 4. Reinforcement rods of specified size and spacing are placed in footings of larger buildings to increase their _____ strength.

_____ 5. The depth of footings must be below the _____ line.

_____ 6. If the soil is stable, form work is not necessary and the concrete can be carefully placed in a _____ of proper width and depth.

_____ 7. When constructing forms, the use of _____ nails ensures easy removal.

_____ 8. _____ are nailed to the top edges of the form to tie them together and keep them from expanding.

_____ 9. A _____ is formed in a footing by pressing 2″ × 4″ lumber into freshly poured concrete.

_____ 10. Sometimes it is necessary to _____ the footing if it is built on sloped land.

_____ 11. Another name for a thickened edge slab is a _____ slab.

_____ 12. Forms for walks and driveways should be built so that water will _____ from the concrete's surface.

_____ 13. When forming curves with ¼" plywood, install it with the grain _____ for easier bending.

_____ 14. _____ the stock also sometimes helps the bending process when forming curves.

Math Problem-Solving

Solve the following math problems.

_____ 1. If a 2 × 6 is 1½" thick, how deep would a saw kerf be if it was to be ⅔ of the thickness?

_____ 2. What is the circumference of a circle whose radius is 15? Answer rounded to two decimal places.

_____ 3. A ¼ bend arc is 90° out of 360°. What is number of degrees in a ⅛ bend arc?

_____ 4. If the inside curve radius of a ¼ bend sidewalk is 5', what is the length of the form needed to make the arc? Answer in decimal feet rounded to two decimal places.

_____ 5. If the outside curve radius of a ¼ bend sidewalk is 9', what is the length of the form needed to make the arc? Answer in feet-inches to the nearest sixteenth.

Discussion

Write your answer(s) on the lines below.

1. There is an old saying among builders, "If you don't have a good foundation, you don't have anything." Explain the reasoning behind this statement.

_____

_____

_____

2. Why is it so important to place a footer below the frost level?

_____

_____

_____

29 Wall and Column Forms

Multiple Choice

Write the letter for the best answer on the line next to the number of the sentence.

_____ 1. To increase the efficiency of forming foundation walls, _____ .
A. the forms are custom-built in place
B. panels and panel systems are used
C. snap ties are eliminated
D. $5\frac{5}{8}''$ plywood is used to construct the walers

_____ 2. A typical _____ size is 2′ wide by 8′ high.
A. waler
B. strongback
C. girder pocket
D. panel

_____ 3. Snap ties are used to _____ .
A. hold the wall forms together at the desired distance
B. support the wall forms against the lateral pressure of the concrete
C. reduce the need for external bracing
D. all of the above

_____ 4. The projecting ends of snap ties are snapped off _____ .
A. prior to placing the concrete
B. inside the concrete after the removal of the forms
C. slightly protruding from the concrete after it is placed
D. when erecting the panels

_____ 5. Walers are _____ .
A. spaced at right angles to the panel frame members
B. always run horizontally
C. constructed of ⅝″ plywood
D. all of the above

_____ 6. The higher a concrete wall, the _____ .
A. more lateral pressure on the top of the form
B. less lateral pressure at the bottom of the form
C. fewer snap ties needed at the bottom of the form
D. greater the lateral pressure on the bottom of the form

_____ 7. To provide a smooth face to the hardened concrete and for easy stripping of the forms, _____ .
A. all panel faces should be oiled or treated with a chemical-releasing agent
B. the panels should be placed horizontally
C. the concrete must not be vibrated
D. it is recommended that the panels not be placed on plates

_____ 8. When concrete walls are to be reinforced, the rebars are _____ .
A. installed after the walers are spaced
B. added after the concrete is placed
C. used instead of snap ties
D. tied in place before the inside panels are erected

_____ 9. A _____ is a thickened portion of the wall added for strength or support for beams.
A. strongback
B. pilaster
C. gusset
D. parapet

_____ 10. Anchor bolts are set in the wall _____ .
A. after the concrete has partially set
B. at the same time as the rebar
C. as soon as the wall is screeded
D. when the concrete is placed

_____ 11. It is important that anchor bolts be _____ .
A. staggered and placed at various heights
B. attached to the rebar
C. set at the correct height and at the specified locations
D. A and B

_____ 12. A _____ is a form that provides an opening in a foundation wall for things such as ducts, pipes, doors, and windows.
A. domeform
B. sleeper
C. keyway
D. buck

Completion

Complete each sentence by inserting the best answer on the line near the number.

_____ 1. A radius can be formed on the corners of a concrete column by fastening _____ molding to the panel edges.

_____ 2. Quarter-round molding can be attached to the panels to form a _____ shape.

_____ 3. By attaching triangular-shaped strips of wood to the edges of a panel, a _____ is formed on the column's corners.

_____ 4. A column may be decorated with flutes by attaching vertical strips of _____ molding spaced on the panel faces.

_____ 5. _____ are often used to provide the face of the column with various textures, such as wood, brick, and stone.

_____ 6. The number and spacing of yokes depends on the _____ of the column.

_____ 7. With column form construction, vertical _____ are installed between the overlapping ends of the yokes.

_____ 8. A concrete _____ consists of manufactured items for concrete form construction.

Math Problem-Solving

Solve the following math problems.

_____ 1. What is the number of cubic feet in a sidewalk that measures 6″ thick × 48″ wide and 56′ long?

_____ 2. How many cubic yards in a slab that measures 4″ × 24′ × 32′? Round answer to nearest one-half cubic yard.

_____ 3. What is the cost of concrete at $75 per cubic yard for a pier that measures 3′ × 9′ × 15′?

_____ 4. How many trucks will be needed to deliver 29 cubic yards of concrete if a 10 yard truck is only allowed to carry 85% of a full load due to weight limits?

_____ 5. A rectangular foundation wall form is erected for a 10″ wall that is 24′ by 36′-6″. How many cubic yards will be needed if 3.5% waste is added? Round answer to nearest one-half cubic yard.

30 Concrete Stair Forms

Multiple Choice

Write the letter for the best answer on the line next to the number of the sentence.

_____ 1. The vertical distance of a step is referred to as _____ .
- A. unit rise
- B. unit run
- C. tread
- D. all of the above

_____ 2. The vertical support under a suspended stair form is called _____ .
- A. horse
- B. shore
- C. post
- D. column

_____ 3. A pitch board allows for _____ .
- A. unit rise
- B. unit run
- C. tread
- D. all of the above

_____ 4. A general thought to keep in mind when forming for concrete stairs is _____ .
- A. use only as many nails as necessary
- B. consider how forms will be removed after concrete placement
- C. clean all adhering concrete from the form material
- D. all the above

_____ 5. How many cubic yards of concrete are necessary for a wall 7" thick, 9' high, and 12' long?
- A. 2.33 cubic yards
- B. 5.25 cubic yards
- C. 63 cubic yards
- D. 756 cubic yards

Completion

Complete each sentence by inserting the best answer on the line near the number.

_____ 1. To conserve on concrete when forming stairs, it may be necessary to lay out the stairs before the _____ is placed.

_____ 2. When forming earth-supported stairs between two existing walls, the _____ and the _____ are laid out on the inside of the existing walls.

_____ 3. When forming stairs, planks are ripped to width to correspond to the height of each _____ .

_____ 4. When forming stairs, it is important to _____ the bottoms of the planks used to form the risers. This permits the mason to trowel the entire edge of the tread.

_____ 5. Riser planks are braced from top to bottom between their ends to keep them from _____ due to the concrete's pressure.

_____ 6. Suspended stairs must be designed and reinforced to support not only their own weight but also the weight of the _____ .

_____ 7. On suspended stairs with open ends, the layout is made on the _____ used for forming the stairs' ends.

_____ 8. Short lengths of narrow boards known as _____ are fastened across joints in forms to strengthen them.

_____ 9. A _____ is a thin piece of plywood that has the width of the tread and the height of the riser laid out on it. It is used to mark the tread and riser locations on the form.

_____ 10. Economical concrete construction depends a great deal on the _____ of form.

Math Problem-Solving

Solve the following math problems.

_____ 1. How many ounces in 12 pounds?

_____ 2. How many pounds to make one ton?

_____ 3. How many tons in one cubic yard of concrete if it weighs 3880 pounds? Round answer to one decimal place.

_____ 4. What is the weight in tons of 8 cubic yards of concrete?

_____ 5. What is the weight in pounds of concrete carried in a wheelbarrow that can hold ½ cubic yard? Round up answer to the nearest pound.

31 Types of Frame Construction

Multiple Choice

Write the letter for the best answer on the line next to the number of the sentence.

_____ 1. The most widely used framing method in residential construction is _____ .
   A. balloon frame construction
   B. platform frame construction
   C. post and beam frame construction
   D. the Arkansas system

_____ 2. A platform frame is easy to construct because _____ .
   A. the second floor joists rest on 1″ × 4″ ribbon
   B. at each level a flat surface is provided to work on
   C. the studs run uninterruptedly the entire height of the building
   D. it uses fewer but larger pieces

_____ 3. Lumber shrinks mostly _____ .
   A. across its width
   B. across its thickness
   C. from end to end
   D. along width and thickness

_____ 4. A raised-heel truss allows more insulation to be installed _____ .
   A. at the eaves
   B. on the roof
   C. in the wall
   D. all of the above

_____ 5. In balloon frame construction, it is important that _____ .
   A. firestops be installed in the walls at several locations
   B. the second floor platform is erected on top of the walls of the first floor
   C. brick or stucco not be used to finish the outside walls
   D. interior design be planned around the supporting roof beam posts

_____ 6. Wood frame construction is used for residential and light commercial construction because _____ .
   A. the cost is usually less than other types
   B. it provides for better insulation
   C. it is very durable and will last indefinitely if properly maintained
   D. all of the above

_____ 7. A post and beam roof is usually _____ .
   A. not insulated
   B. insulated on top of the deck
   C. insulated below the deck
   D. more labor- and material-consuming than conventional roof framing

_____ 8. House depths that are not evenly divided by four tend to _____ .
   A. conserve material
   B. waste material
   C. save on labor costs
   D. A and C

_____ 9. Reducing the clear span of floor joists _____ .
   A. makes it necessary to use larger joists
   B. can be accomplished by using narrower sill plates
   C. can make it possible to use smaller-sized joists
   D. can always result in higher costs if wider sill plates must be used

_____ 10. The material used between joists to stiffen them is called _____ .
   A. sill sealer
   B. bridging
   C. firestops
   D. subfloor

Math Problem-Solving

Solve the following math problems.

_____ 1. If a wet 2 × 6 board was seen to expand in thickness ⅛″, by what percent did it increase?

_____ 2. What is the perimeter of a rectangular building that measures 32′ × 64′?

_____ 3. What is the perimeter of the foundation shown in the figure A?

_____ 4. What is the smallest number of 12′ long sill pieces would be needed for the foundation of A?

_____ 5. How many rolls of sill sealer would be needed for the foundation in figure A if each roll was 50′ long?

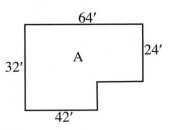

Discussion

Write your answer(s) on the lines below.

1. Explain in detail what occurs when building dimensions are divisible by four.

_____

_____

2. Given a choice between platform, balloon, or post and beam construction in a home you would be building for yourself, which would you choose and why?

_____

_____

32 Layout and Construction of the Floor Frame

Multiple Choice

Write the letter for the best answer on the line next to the number of the sentence.

_____ 1. Heavy beams that support the inner ends of the floor joists are called _____ .
A. sill plates
B. girders
C. subflooring
D. bridging

_____ 2. If the set of prints does not specify the kind or size of structural component, _____ .
A. the job foreman may have to estimate this data
B. always oversize the material to be safe
C. have the design checked by a professional engineer
D. consult with an insurance expert

_____ 3. Pockets should be large enough to provide at least a _____ bearing for the girder.
A. 12″
B. 1½″
C. 8″
D. 4″

_____ 4. Anchor bolts are used to attach the _____ to the foundation.
A. floor joists
B. sill plate
C. girder
D. all of the above

_____ 5. When tightening the nuts on anchor bolts, be sure to _____ .
A. replace the washers first
B. not overtighten them
C. tighten them as tight as possible
D. A and B

_____ 6. In conventional framing, floor joists are usually placed _____ .
A. 16″ on center
B. prior to the girder
C. 18″ on center
D. below the sill plate

_____ 7. Notches in the bottom or top of sawn lumber floor joists should not _____ .
A. exceed ⅜ the joist depth
B. be in the middle third of the joist span
C. be in the first third of the joist span
D. exceed ½ the joist depth

_____ 8. When measuring for floor joist layout, it is best to _____ .
   A. mark the joist locations on top of the foundation
   B. measure and mark each joist individually
   C. have the ends of the plywood fall directly on the edge of the joists
   D. use a tape stretched along the length of the building

_____ 9. Joists are installed with the _____ .
   A. crown down
   B. lap-spiked together over the sill plate
   C. crown up
   D. A and B

_____ 10. When installing plywood subflooring _____ .
   A. leave a ¹⁄₁₆″ space at all panel end joints
   B. leave an ⅛″ between panel edges
   C. all end joints are made over joists
   D. A, B, and C

Completion

Complete each sentence by inserting the best answer on the line near the number.

_____ 1. Sill plates lie directly on the foundation wall and provide bearing for the _____ .

_____ 2. If joists are lapped over the girder, the minimum amount of lap is _____ inches.

_____ 3. Holes bored in joists for piping or wiring should not be larger than _____ of the joist depth.

_____ 4. Shortened floor joists at the ends of floor openings are called _____ joists.

_____ 5. It is important that the floor joist layout permits the plywood sheets to fall directly on the _____ of the joists.

_____ 6. The area of a rectangular floor is determined by multiplying the length by its _____ .

_____ 7. To determine the number of rated panels of subflooring required, divide the floor area by _____ .

_____ 8. A one-story home with the dimensions of 28′ × 48′ would require _____ rated panels for the subfloor.

_____ 9. To determine the total linear feet of wood cross-bridging needed, multiply the length of the building by _____ for every row of bridging.

Identification: Floor Frame

Identify each term, and write the letter of the best answer on the line next to each number.

_____ 1. girder

_____ 2. floor joist

_____ 3. rim joists

_____ 4. treated sill

_____ 5. foundation wall

_____ 6. column footing

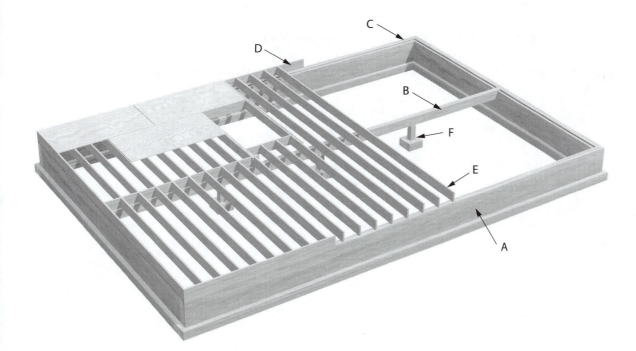

Math Problem-Solving

Solve the following math problems.

_____ 1. What is the post spacing if seven posts are to be installed under a girder that is 57'-9" long which also rest at either end on the foundation? Answer in inches to the nearest eighth.

_____ 2. What is the butt seam separation of a built-up girder lams if the 1/5 rule is used and the post spacing is 8'-4"? Answer in inches.

The following questions pertain to a rectangular floor frame that is 32' × 52'

_____ 3. How many 2 × 6-16' sills would be needed?

_____ 4. How many joists would be need if they lap at the girder and four extra are needed for doubling?

_____ 5. How many sheets of 4' × 8' plywood to sheath the deck?

_____ 6. How many lineal feet of 1" × 3" stock is needed for wood cross bridging?

33 Construction to Prevent Termite Attack

Multiple Choice

Write the letter for the best answer on the line next to the number of the sentence.

_____ 1. For effective treatment against dry wood termites, it is usually necessary to

_____ .

    A. install earth-to-wood termite barriers, and chemically treat the foundation and soil
    B. tent-fumigate the whole home with a toxic gas
    C. eliminate any moisture that is reaching the wood and allow the wood to dry out
    D. spray the infested area with over-the-counter insecticides

_____ 2. The most destructive species of termite is the _____ .

    A. dry wood termite
    B. damp wood termite
    C. subterranean termite
    D. northern pine termite

_____ 3. Protection against subterranean termites should be considered _____ .

    A. at the first sign of an infestation
    B. when planning and during construction of a building
    C. prior to occupation of the building
    D. after the elimination of the queen

_____ 4. Prevention of termite attacks is based on _____ .

    A. keeping the wood dry
    B. making it as difficult as possible for termites to get to the wood
    C. chemical treatment of the soil
    D. all of the above

_____ 5. In crawl spaces, clearance between the ground and the bottom of the floor joists
should be at least _____ .

    A. 8″
    B. 12″
    C. 14″
    D. 18″

_____ 6. The _____ foundation provides the home with the best protection against termites.

    A. slab-on grade
    B. monolithic slab
    C. two core concrete block
    D. independent slab

_____ 7. Wall siding should not extend more than _____ below the top of the foundation wall.

    A. 2″
    B. 4″
    C. 6″
    D. 8″

_____ 8. Research has shown that due to _____ termite shields have not been effective in preventing infestations.
   A. improper installation
   B. the failure to frequently inspect for signs of shelter tubes
   C. the shields' rapid deterioration
   D. A and B

_____ 9. To safely work with pressure treated lumber, it is important _____ .
   A. to wear eye protection and a dust mask when sawing or machining it
   B. upon completion of the work to wash your hands before eating and drinking
   C. not to burn the leftover scraps
   D. all of the above

_____ 10. Clothing that accumulates sawdust from pressure treated wood should be _____ .
   A. disposed of
   B. treated with a disinfectant
   C. laundered separately from other clothing and before reuse
   D. washed in boiling water, then dried outdoors

Completion

Complete each sentence by inserting the best answer on the line near the number.

_____ 1. _____ termites tend to cause less damage to buildings than other types of termites.

_____ 2. Discovery of a leak sometimes reveals a _____ termite infestation.

_____ 3. All wood scraps should be _____ from the area before backfilling around the foundation.

_____ 4. A pit in the ground filled with stones used to absorb drainage water is called a _____ .

_____ 5. Cracks as little as _____ of an inch permit the passage of termites.

_____ 6. Wall siding should be at least _____ above the finish grade.

_____ 7. _____ grade pressure treated lumber is used for sill plates, joists, girders, and decks.

_____ 8. Termites will not usually eat treated wood; however, they will _____ over it to reach wood that is not treated.

Math Problem-Solving

Solve the following math problems.

_____ 1. What is the sum of 6¾₆″, 5⅜″ and 10¹⁵⁄₁₆″?

_____ 2. Subtract 11¹¹⁄₁₆″ from 20⅝″.

_____ 3. Convert 0.59″ to the nearest sixteenth of an inch.

_____ 4. What is one-fifth of 14′-6″ in feet-inches to the nearest sixteenth?

_____ 5. Convert 104.65′ to feet-inches to the nearest sixteenth.

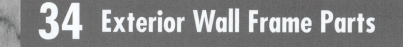

34 Exterior Wall Frame Parts

Multiple Choice

Write the letter for the best answer on the line next to the number of the sentence.

_____ 1. It is important for the carpentry student to be able to know the _____ of the
different parts of the wall frame.
   A. names
   B. functions
   C. locations
   D. all of the above

_____ 2. Names given to different parts of a structure _____ .
   A. are the same nationwide
   B. may differ with the geographical area
   C. always include the size of the framing member
   D. change if engineered lumber is used

_____ 3. The bottom horizontal member of a wall frame is called the _____ .
   A. sole plate
   B. top plate
   C. double plate
   D. sill plate

_____ 4. Vertical members of the wall frame that run full length between the plates are known as
_____ .
   A. ribbons
   B. trimmers
   C. braces
   D. studs

_____ 5. It is necessary that headers be _____ .
   A. installed wherever interior partitions meet exterior walls
   B. left in until the wall section is erected
   C. strong enough to support the load over the opening
   D. placed under the rough sill

_____ 6. When headers are built to fit a 4″ wall, flush on both wall surfaces, they will often consist of
_____ .
   A. two pieces of 2″ lumber
   B. three pieces of 2″ lumber
   C. two pieces of 2″ lumber with ½″ plywood or strand board sandwiched in between
   D. two pieces of 2″ lumber with ¾″ plywood or strand board sandwiched in between

_____ 7. Some carpenters prefer to double the rough sills to _____ .
   A. bear the weight over the window
   B. cut down on insulation
   C. provide more surface to nail window trim
   D. fasten the drywall to

_____ 8. Shortened studs that carry the weight of the header are called _____ .
   A. trimmers
   B. door jacks
   C. braces
   D. A and B

_____ 9. Ribbons are horizontal members that support the second floor _____ .
   A. partition intersections
   B. joists in balloon construction
   C. corner posts
   D. headers

_____ 10. Generally, corner bracing is not needed if _____ is used on the corners.
   A. softboard
   B. Celotex
   C. insulated board sheathing
   D. rated panel wall sheathing

Completion

Complete each sentence by inserting the best answer on the line near the number.

_____ 1. Studs are usually _____ , unless 6″ insulation is desired in the exterior wall.

_____ 2. Studs are usually spaced _____ or on center.

_____ 3. The depth of the header depends on the _____ of the opening.

_____ 4. Headers in _____ walls may be made up of three pieces of 2″ lumber and two pieces of ½″ plywood or strand board.

_____ 5. The use of _____ lumber for headers permits the spanning of wide openings that otherwise might need additional support.

_____ 6. Rough sills are members used to form the bottom of a _____ opening.

_____ 7. Corner posts may be constructed of _____ 2″ × 4″s.

_____ 8. Ribbons are usually made of _____ stock.

_____ 9. When needed, wall corners are braced using 1″ × 4″s that are set into the face of the studs, top plate, and sole plate, and run _____ .

Identification: Exterior Wall Frame Parts

Identify each term, and write the letter of the correct answer on the line next to each number.

_____ 1. corner post

_____ 2. top plate

_____ 3. sole plate

_____ 4. OC stud

_____ 5. cripple stud

_____ 6. cut-in bracing

_____ 7. partition stud assembly

_____ 8. trimmer

_____ 9. rough sill

_____ 10. ribbon

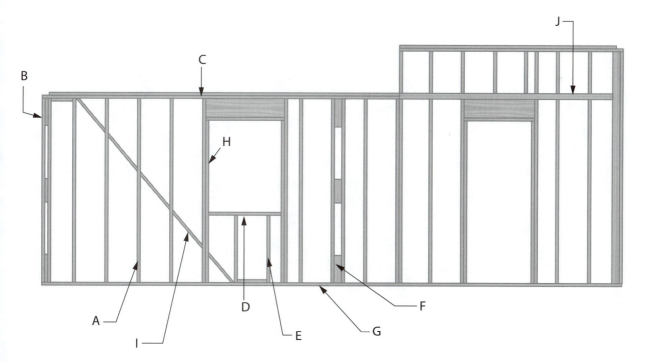

Math Problem-Solving

Solve the following math problems.

_____ 1. Multiply ⅙ × ⅔ and express answer as a fraction in lowest terms.

_____ 2. Divide ⅘ by ⅔ and express answer as a fraction in lowest terms.

_____ 3. How many pounds would 250 studs weigh if each weighs 4³⁄₁₆ pounds? Round up to the nearest pound.

_____ 4. How many 2 × 10 headers can be purchased with $150, if each is priced at $2.30?

_____ 5. How many pieces of blocking, 1⅓′ long, may be cut from a board that is 14′ long?

35 Framing the Exterior Wall

Multiple Choice

Write the letter for the best answer on the line next to the number of the sentence.

_____ 1. To determine stud length, the carpenter must know the _____ .
   A. thickness of the finished floor
   B. thickness of the ceiling below the ceiling joists
   C. height from the finished floor to the finished ceiling
   D. all of the above

_____ 2. The carpenter must determine _____ sizes from information contained in the door and window schedules.
   A. header
   B. corner post
   C. rough opening
   D. A and B

_____ 3. When laying out a rough opening, a clearance of _____ is usually maintained between the door frame and the rough opening.
   A. ¼″
   B. ½″
   C. ¾″
   D. 1″

_____ 4. The sides and tops of a door frame are called _____ .
   A. stops
   B. rabbets
   C. jambs
   D. door sets

_____ 5. Interior partitions on a set of prints are usually _____ .
   A. dimensioned to their centerlines
   B. shown on the plot plan
   C. eliminated
   D. represented by dotted lines on the floor plan

_____ 6. When laying out studs, _____ .
   A. vary their spacing between 8″ and 10″
   B. keep the studs directly in line with the joists below
   C. mark their location on the sill plate
   D. all of the above

_____ 7. Prior to bracing the end sections of a wall, an easy and effective way to check the wall section for square is to _____ .

    A. use a framing square

    B. measure the corner posts to see if they are the same

    C. measure the wall from corner to corner both ways to see if they are the same

    D. check the window openings with a combination square

_____ 8. For accurate plumbing of corner posts on a windy day use _____ .

    A. a builders level

    B. a plumb bob

    C. a 6′ level with accessory aluminum blocks attached to each end

    D. all of the above

_____ 9. A job-built combination wall aligner and brace is called a/an _____ .

    A. spring brace

    B. wall jack

    C. aligner strut

    D. partition brace

_____ 10. Softboard sheathing is fastened to the wall using _____ nails.

    A. 6d

    B. 12d

    C. duplex

    D. roofing

Completion

Complete each sentence by inserting the best answer on the line near the number.

_____ 1. A _____ is an opening framed in the wall in which to install doors and windows.

_____ 2. The bottom member of a door frame is called a _____ .

_____ 3. Consulting the manufacturers' _____ is the best way to determine window rough openings.

_____ 4. The first step in laying out wall openings is to consult the set of prints to find the _____ dimension of all the openings.

_____ 5. To avoid problems when installing finish work, it is important that all edges of frame members be kept _____ wherever they join each other.

_____ 6. All studs in a wall should have their crowned edges facing in the _____ direction.

_____ 7. When applying temporary bracing to an erected wall, use _____ nails or drive the nails only partway into the lumber.

_____ 8. Partitions that carry a load are referred to as _____ partitions.

_____ 9. Wall _____ covers the exterior walls.

_____ 10. When estimating material for an exterior wall with a layout that is 16″ on center, figure _____ stud(s) for each linear foot.

Math Problem-Solving

Solve the following math problems.

_____ 1. What is the true door width if ³⁄₃₂″ is planed off a 3′-0″ door? Answer in terms of feet-inches to nearest 32nd.

_____ 2. What is the true door height if ³⁄₃₂″ is planed off a 6′-8″ door? Answer in terms of feet-inches to nearest thirty-second.

Use the following information to solve the remaining questions. Door is 3′-0″ × 6′-8″, jambs are ¾″ thick, finished floor is a total of 1″ thick, finished ceiling is ½″ drywall, and finished floor to ceiling height is to be 9′-0″, normal construction with dimension lumber.

_____ 3. What is the master stud length? Answer in terms of inches.

_____ 4. What is the header height from subfloor? Answer in terms of inches.

_____ 5. What is the jack stud length? Answer in terms of inches.

36 Interior Partitions and Ceiling Joists

Multiple Choice

Write the letter for the best answer on the line next to the number of the sentence.

_____ 1. Because of the added stress on bearing walls, it is required that _____ .
   A. the top plate be doubled
   B. the sole plate be doubled
   C. they be erected after the roof is on
   D. they be erected in a manner different than exterior walls

_____ 2. Roof trusses eliminate the need for bearing partitions because _____ .
   A. they distribute all the weight on the interior walls
   B. the girder carries the weight
   C. they transmit the weight to the exterior walls
   D. they weigh less than conventional roof framing

_____ 3. Rough opening widths for interior doors are equal to the _____ .
   A. door width plus 1″ and the thickness of the finish floor
   B. door width plus 1″ and twice the thickness of the door stop
   C. door width plus 1″ and twice the thickness of the door jamb
   D. width of the door plus 1¾″

_____ 4. At exterior walls, ceiling joists are placed so that _____ .
   A. the sole plate is attached to them
   B. the rafters can be attached to their sides
   C. they will support the roof trusses until they are set into place
   D. A and C

_____ 5. When ceiling joists are installed in line, _____ .
   A. their ends must butt together at the center line of the bearing partition
   B. a scab must be installed at the splice
   C. the exterior wall location for either side of the span is on the same side of the rafter
   D. all of the above

_____ 6. The ends of ceiling joists on the exterior wall must be cut _____ .
   A. if additional head room is needed
   B. slightly above the rafter at the same angle as the roof's slope
   C. flush or slightly below the top edge of the rafter
   D. only if the roof has a very steep pitch

_____ 7. When low-pitched hip roofs are used, _____ .
   A. roof trusses are a necessity
   B. ceiling joists must be doubled
   C. stub joists must be installed
   D. ceiling joists may be eliminated

_____ 8. Ceiling joists on each end of the building are placed so that _____ .
   A.   the outside face is flush with the inside of the wall
   B.   the outside face is flush with the outside of the wall
   C.   there is a 1½″ gap between the joists and outside wall
   D.   there is access for the electrician to run wiring through them

_____ 9. The top plate of a bearing partition _____ .
   A.   laps the plate of the exterior wall
   B.   is a single member
   C.   butts the top plate of the exterior wall
   D.   is applied after the ceiling joists are installed

_____ 10. The member added to the top of ceiling joints to add stiffness is called _____ .
   A.   ribband
   B.   strongback
   C.   blocking
   D.   A and B

Completion

Complete each sentence by inserting the best answer on the line near the number.

_____ 1. Ceiling joists tie the exterior side walls together and provide a base for the ceiling _____ .

_____ 2. Size and spacing of ceiling joists are determined by the _____ .

_____ 3. _____ prevent the roof from exerting outward pressure on the wall, which would cause the walls to spread.

_____ 4. When joists are installed, be sure the crowned edges are pointed _____ .

_____ 5. When cutting the taper on the ends of ceiling joists, be sure the taper length does not exceed _____ times the depth of the member.

_____ 6. A _____ is a small opening in ceiling joists that allows access to the attic.

_____ 7. Joists are to be toenailed into the plates with at least two _____ penny nails.

_____ 8. Ceiling joists may be installed butting a girder with each joist supported by a _____ .

_____ 9. A non-load bearing header may be built with _____ .

_____ 10. Bathroom and kitchen walls sometimes must be made thicker to accommodate _____ later installed in those walls.

_____ 11. A rough opening width of _____″ is needed for a 32″ interior door, if the jamb stock is ¾″ thick.

Math Problem-Solving

Solve the following math problems.

The following questions pertain to the following information: rectangular building measuring 30′ × 52′, framed with 16″ OC studs.

_____ 1. Estimate how many exterior studs would be needed.

_____ 2. How many plates, 16′ long, would be needed?

_____ 3. How many plates, 12′ long, would be needed?

_____ 4. How many sheets of sheathing would be needed? Neglect covering the box header and any waste factor.

_____ 5. How many sheets of sheathing would be needed if the 1′ wide box header was also being covered and 3% was added for waste?

37 Backing and Blocking

Multiple Choice

Write the letter for the best answer on the line next to the number of the sentence.

_____ 1. Another common term for backing is _____ .
   A. cripple
   B. blocking
   C. nailer
   D. all of the above

_____ 2. It is a good idea to install baseboard backing _____ .
   A. in every corner of every room
   B. on either side of a window openings
   C. only in rooms with exterior walls
   D. all of the above

_____ 3. Backing is needed in kitchens for _____ .
   A. base cabinets
   B. refrigerators
   C. faucets
   D. all the above

_____ 4. To speed up the nailing of horizontal wall blocking _____ .
   A. use a pneumatic nailer
   B. pre-drive toe nails before placing block between studs
   C. use a wood scrap as a temporary post
   D. all of the above

Completion

Complete each sentence by inserting the best answer on the line near the number.

_____ 1. A short block of lumber that is installed in floor, wall, and ceiling cavities to provide fastening for various fixtures and parts is known as _____ .

_____ 2. The placement of blocking and backing is not usually found in a set of _____ .

_____ 3. Wall blocking is needed at the _____ edges of wall sheathing panels permanently exposed to the weather.

_____ 4. Fire Blocking is required between studs in walls over _____ high.

_____ 5. It is easier to install blocking in a _____ line than in a straight one.

_____ 6. Blocking is not necessary under joints in the subfloor when the panel edges are _____ .

_____ 7. _____ type blocking is used to anchor non-load bearing top plates that run parallel to the joists.

_____ 8. When installing a line of blocking it is best to periodically _____ a block to make sure the joists or studs remain straight.

_____ 9. The room that needs the most number of blocking pieces is the _____ .

_____ 10. A _____ nail may be used when installing blocking in a straight line.

Identification: Blocking and Backing

Identify each term, and write the letter of the best answer on the line next to each number.

_____ 1. faucet backing

_____ 2. tub support backing

_____ 3. showerhead backing

_____ 4. outlet backing

_____ 5. shower curtain rod backing

_____ 6. lavatory backing

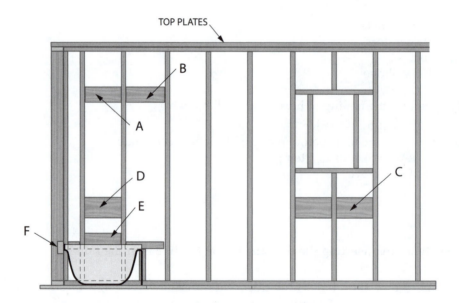

SECTION THROUGH BATHTUB

Math Problem-Solving

Solve the following math problems.

_____ 1. How many lineal feet of fireblocking are needed over a girder where the floor system measures 28′ × 64′?

_____ 2. Neglecting window and door openings, how many lineal feet of blocking would be needed for 15′ of upper and lower kitchen cabinets where studs are 16″ OC?

_____ 3. Neglecting window and door openings, how many lineal feet of blocking would be needed for 17′ of upper and lower kitchen cabinets where studs are 24″ OC?

_____ 4. What is the length, in inches, of a header needed over two doors that are 3′-0″ wide?

_____ 5. What is the actual distance between 16″ OC studs framed with dimension lumber?

38 Steel Framing

Multiple Choice

Write the letter for the best answer on the line next to the number of the sentence.

_____ 1. The components of steel framing include _____ .

    A. studs, plates, and runners
    B. tracks, plates, and channels
    C. channels, studs, and tracks
    D. runners, tracks, and studs

_____ 2. Steel is called galvanized when it is coated with _____ .

    A. oil
    B. zinc
    C. stainless steel
    D. paint

_____ 3. The term in steel framing that is similar to width in wood is _____ .

    A. thickness
    B. flange
    C. depth
    D. lip size

_____ 4. The size of steel designated for structural framing is _____ .

    A. 25 to 20 gauge
    B. 18 to 33 mils
    C. 0.5 to 0.8 mm
    D. 43 to 97 mils

_____ 5. Hat tract is also referred to as _____ .

    A. metal furring
    B. plate material
    C. runner
    D. all of the above

_____ 6. Steel studs are best cut with a _____ .

    A. pair of tin snips
    B. compound miter saw
    C. chop box
    D. hack saw

_____ 7. Steel framing is similar to wood framing in that _____ .

    A. walls are preassembled and then stood up
    B. layout is done on horizontal members for OC studs
    C. each stud receives two nails
    D. all of the above

_____ 8. Wood is sometimes added to steel framing to _____.
   A. increase a track's load bearing capacity
   B. provide a nailer for trim
   C. provide fire resistance
   D. all of the above

_____ 9. Punch outs in steel studs are there to _____.
   A. provide holes to run wires
   B. make the stud lighter
   C. improve the stiffness of stud
   D. make attaching studs to tracks easier

_____ 10. When two steel studs are interlocked, one inside the other, it is called _____.
   A. stiffening
   B. strengthening
   C. nesting
   D. locking

Completion

Complete each sentence by inserting the best answer on the line near the number.

_____ 1. All steel framing members are coated with material that resists _____ .

_____ 2. Studs for interior nonload-bearing applications are manufactured from 18-, 27-, and _____ -mil steel.

_____ 3. Pipes and conduit can be ran through punchouts that are located at intervals in the studs _____ .

_____ 4. The top and bottom horizontal members of a steel frame wall are called _____ .

_____ 5. Steel channels are used in suspended ceilings and for _____ of walls.

_____ 6. Maximum spacing for metal furring channels is _____ inches on center.

_____ 7. CRC in steel framing stands for _____ .

_____ 8. Heavy gauge steel framing requires fasteners that have _____ points.

_____ 9. During installation of a one-piece door buck the _____ of the king studs are left unfastened until after buck is inserted.

_____ 10. Metal furring may be used as on _____ and _____ to support drywall.

Math Problem-Solving

Solve the following math problems.

_____   1. What should be the actual length, in inches, of steel stud in a wall that is 8'-0" tall?

_____   2. What is the thickness, to the nearest sixteenth of an inch, of 54 mil thick CRC if one mil equals 1/1000"?

_____   3. Convert 6.07' to feet-inches to the nearest sixteenth.

_____   4. What is the minimum number of 16" OC studs would actually be needed for a 14'-8" long wall section with no opening?

_____   5. How many pieces of track would be needed for a 14'-8" wall section with no openings?

Discussion

Write your answer(s) on the lines below.

1. What are some of the factors that could lead to steel framing members being used more frequently in the future?

_____

_____

_____

2. What usually determines whether specialists in light steel or carpenters do the framing?

_____

_____

_____

3. Over an extended period of time, which type of stud—steel or wood—is more environmentally favorable? Explain your answer.

_____

_____

_____

39 Wood, Metal, and Pump Jack Scaffolds

Multiple Choice

Write the letter for the best answer on the line next to the number of the sentence.

_____ 1. All scaffolds must be capable of supporting without failure at least _____ .
    A. the maximum intended load
    B. two times the maximum intended load
    C. three times the maximum intended load
    D. four times the maximum intended load

_____ 2. On metal frame scaffolding, _____ should never be used as a means of access or egress.
    A. cross braces
    B. end frames
    C. extension ladders
    D. A and B

_____ 3. Vertical members of a scaffold are called _____ .
    A. ledgers
    B. bearers
    C. braces
    D. poles

_____ 4. To prevent excessive checking, scaffold planks should _____ .
    A. be painted
    B. not exceed 6' in length
    C. have their ends banded with steel
    D. be replaced yearly

_____ 5. All scaffold planks must _____ .
    A. be scaffold grade or its equivalent
    B. be laid with their edges close together
    C. not overhang the bearer by more than 12"
    D. all of the above

_____ 6. On scaffolds that are more than ten feet high, _____ are installed on all open sides.
    A. bearers
    B. outriggers
    C. bucks
    D. guardrails

_____ 7. Pipe scaffolds should be set _____ .
    A. up tight against the wall
    B. as far from the wall as the worker can comfortably reach
    C. as close to the wall as is possible without interfering with the work
    D. without braces on the inside face on medium duty scaffolds

_____ 8. Scaffolding should be erected _____.
   A. by an OSHA inspector
   B. under supervision of a competent person
   C. by a subcontractor
   D. using new equipment

_____ 9. With metal scaffolding, _____.
   A. only one set of braces is needed
   B. braces must be forced on to fit correctly
   C. each section consists of two end pieces and two folding braces
   D. more time is needed to erect it, due to its difficulty to work with

_____ 10. Pump jack scaffolds should not be used when _____.
   A. the working level exceeds 500 lbs.
   B. more than two people will work on the scaffold
   C. the poles exceed 30' in height
   D. all of the above

Completion

Complete each sentence by inserting the best answer on the line near the number.

_____ 1. The _____ is the one responsible for the safe erection of scaffolding.

_____ 2. The work area of a scaffold must be _____ between the outside uprights and the guardrail system.

_____ 3. _____ are diagonal members that stiffen the scaffolding and prevent the poles from moving or buckling.

_____ 4. The top guardrail should be _____ inches above the working platform.

_____ 5. Scaffold _____ are allowed to do their work under less restrictive safety requirements than scaffold users.

_____ 6. Scaffold planks should be of equal lengths so that the ends are _____ with each other.

_____ 7. Mobile scaffold towers can easily _____ if used incorrectly.

_____ 8. When erecting metal scaffolding, always level it until the _____ fit easily.

_____ 9. Casters for mobile scaffolding must be able to support _____ times the maximum intended load.

_____ 10. On pump jack scaffold poles, the braces must be installed at vertical intervals not exceeding _____ feet.

144

Identification: Scaffold Parts

Identify each term, and write the letter of the correct answer on the line next to each number.

_____ 1. scaffold pole

_____ 2. bearer

_____ 3. top rail

_____ 4. mid-rail

_____ 5. toe board

_____ 6. ledger

_____ 7. scaffold plank

_____ 8. wall ledger

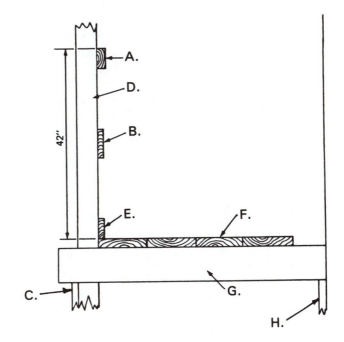

Math Problem-Solving

Solve the following math problems.

_____ 1. How many pounds must a scaffold be built to hold if two 200 pound workers and 200 pounds of material are expected to be supported?

_____ 2. Round 23.45389 to the nearest hundredth.

_____ 3. If a final calculation reveals 14.1546 rated panels are needed to sheath a subfloor, how many should be ordered?

_____ 4. What is the amount of sales tax, at 8%, on a $15,945.00 order of lumber?

_____ 5. If 1" is equal to 2.54 cm, how many centimeters (cm) in 6"?

Name_____   Date _____

40 Brackets, Horses, and Ladders

Multiple Choice

Write the letter for the best answer on the line next to the number of the sentence.

_____ 1. Roof jacks are used to _____ .
   A. raise the roof when remodeling roof
   B. support workers and materials during roofing
   C. support siding material during installation
   D. all of the above

_____ 2. Toe boards _____ .
   A. are required on scaffolding
   B. help prevent tools from falling
   C. protect shingles
   D. all of the above

_____ 3. If a 28' extension ladder is placed against the side of a tall building, the bottom of the ladder should be about _____ from the wall
   A. 3'
   B. 4'
   C. 7'
   D. 9'

_____ 4. Extension ladders used to access the roof must _____ .
   A. extend 3' above roof edge
   B. have feet placed on firm footing
   C. be protected from sideways movement
   D. all of the above

_____ 5. Ladder brackets are installed to _____ .
   A. support scaffold planks
   B. brace ladder from undue flexing
   C. brace ladder from sliding sideways
   D. support paint cans

_____ 6. Rungs of job-made ladders should be dadoed into the rail _____ .
   A. ⅜"
   B. between ⅜" to 1½"
   C. ⅜" and have cleats installed between rungs
   D. all of the above

Completion

Complete each sentence by inserting the best answer on the line near the number.

_____ 1. Roof brackets are usually required whenever the roof has more than a _____-inch vertical rise per horizontal unit of run.

_____ 2. Roof brackets should be placed at about _____-foot horizontal intervals.

_____ 3. When nailing roof brackets, use three 12d common nails driven all the way; try to get at least one nail into a _____ .

_____ 4. A _____ is a low working platform supported by a bearer with spreading legs at each end.

_____ 5. For light duty work, horses and trestle jacks should not be spaced more than _____ feet apart.

_____ 6. If a horse scaffold is arranged in tiers, no more than _____ tiers should be used.

_____ 7. The base of an extension ladder should be held a distance out from the wall equal to _____ the ladder's vertical height.

_____ 8. When used to reach a roof or a working platform, the top of the ladder must extend at least _____ feet above the top support.

_____ 9. Always be careful of overhead electric lines, especially when using a/an _____ ladder.

_____ 10. Metal brackets installed on ladders to hold scaffold planks are called _____ .

_____ 11. The ladder rungs of a job-built ladder must be _____ the ladder rails.

_____ 12. Safety is the responsibility of _____ on the job.

Math Problem-Solving

Solve the following math problems using the diagram shown.

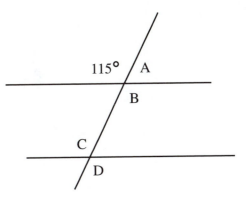

_____ 1. What is the measure of angle A?

_____ 2. What is the measure of angle B?

_____ 3. What is the measure of angle C?

_____ 4. What is the measure of angle D?

_____ 5. What is the sum of 116° 35′ 40″ and 63° 24′ 20″?

Discussion

Write your answer(s) on the lines below.

1. What are some of the things those responsible for erecting scaffolds must be aware of?

   _____

   _____

   _____

2. Do you think it would be advisable for construction firms to use professional scaffold erectors? If so, when and why?

   _____

   _____

   _____

41 Roof Types and Terms

Multiple Choice

Write the letter for the best answer on the line next to the number of the sentence.

_____ 1. The most common roof type is called the ___*B*___ .

    A. common roof
    B. gable roof
    C. intersecting roof
    D. hip roof

_____ 2. The tool used to lay out rafters is called ___*D*___ .

    A. speed square
    B. framing square
    C. rafter square
    D. all of the above

_____ 3. If the slope for a roof is 8 on 12, unit rise of a common rafter is ___*A*___ .

    A. 8
    B. 12
    C. 4
    D. 96

_____ 4. If the slope of a roof is 10 on 12, the unit run of a common rafter is ___*B*___ .

    A. 10
    B. 12
    C. 2
    D. 17

_____ 5. If the span of a building is 32' and the slope is 11 on 12, the total run is ___*C*___ .

    A. 11
    B. 12
    C. 16
    D. 32

Matching

Write the letter for the best answer on the line near the number to which it corresponds.

_E_ 1. gable roof   A. the longer side is 2″ wide

_F_ 2. shed roof   B. uppermost horizontal line of the roof

_I_ 3. hip roof   C. usually the width of the building

_N_ 4. gambrel roof   D. inches of rise per foot of run

_C_ 5. span   E. two sloping roofs meeting at the top

_G_ 6. rafter   F. slopes in one direction only

_M_ 7. total run   G. sloping member of the roof frame

_B_ 8. ridge   H. vertical when the rafter is in position

_J_ 9. total rise   I. slopes upward from all walls of the building to the top

_L_ 10. line length   J. vertical distance that the rafter rises

_O_ 11. pitch   K. horizontal when the rafter is in position

_D_ 12. slope of the roof   L. gives no consideration to the thickness of the stock

_H_ 13. plumb cut   M. horizontal travel of the rafter

_K_ 14. level cut   N. a variation of the gable roof

_A_ 15. framing square   O. a fractional ratio of rise to span

Identification: Roofs

Referring to the following figures, identify each term and write the letter of the best answer on the line next to each number.

___D___ 1. gable roof

___E___ 2. shed roof

___F___ 3. hip roof

___A___ 4. intersecting roof

___B___ 5. gambrel roof

___C___ 6. mansard roof

A.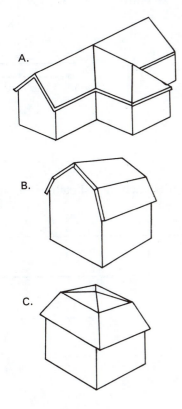

B.

C.

D.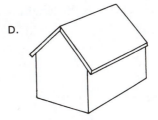

E.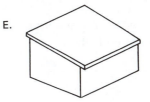

F.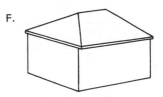

Math Problem-Solving

Solve the following math problems.

_____ 1. Evaluate the solution of $\sqrt{(8^2 + 12^2)}$ to the nearest hundredth.

_____ 2. Evaluate the solution of $\sqrt{(6^2 + 16.9705^2)}$ to the nearest hundredth.

_____ 3. Given the legs of a right triangle are 6 and 8, what is the measure of the longest side?

_____ 4. What is the hypotenuse of a right triangle with sides measuring 13'-9" and 8'-2" in terms of feet-inches to the nearest sixteenth?

_____ 5. What is the length of side A in the figure shown?

Discussion

Write your answer on the lines below each instruction.

1. Describe the line length method of calculating rough rafter lengths. Include a sketch of the process in your description.

   _____

   _____

   _____

Name_____   Date _____

42 Gable and Gambrel Roofs

Multiple Choice

Write the letter for the best answer on the line next to the number of the sentence.

_____   1. During the layout of a common rafter at least ___C___ plumb lines are drawn.
   A. 1
   B. 2
   C. 3
   D. 7

_____   2. The bird's mouth is ___D___ .
   A. made with two lines
   B. cut to fit on top of wall plate
   C. near the bottom end of the rafter
   D. all of the above

_____   3. If a building has a span of 32′ and a rafter slope of 5 on 12, the unit length of the rafter is ___C___ .
   A. 5
   B. 12
   C. 13
   D. 16

_____   4. If a building has a span of 32′ and a rafter slope of 5 on 12, the line length of the rafter is ___D___ .
   A. 90″
   B. 12′
   C. 16′
   D. 208″

_____   5. Framing square gauges are ___D___ .
   A. used to hold rafter square at the same angle during layout
   B. held against the rafter edge during layout
   C. attached to the framing square
   D. all of the above

_____   6. If a building has a projection of 18″, the number of step-offs for laying out the overhang is said to be ___B___ .
   A. 1
   B. 1½
   C. 2
   D. 18

_____   7. The term used to indicate the length of a rafter with a deduction added for the half-thickness of the ridge is ___A___ .
   A. actual length
   B. theoretical line length
   C. rafter length
   D. all of the above

_____ 8. Truss joist rafters must have _____*A*_____ installed at the bird's mouth.

    A. web stiffeners

    B. LVLs

    C. joist hangers

    D. all of the above

_____ 9. If a building is 30' long with a rafter slope of 4 on 12, the recommended collar tie length is _____*D*_____ .

    A. 4'

    B. 5'

    C. 8'

    D. 10'

_____ 10. Lookouts support _____*A*_____ .

    A. rake rafters

    B. gable studs

    C. gambrel rafters

    D. collar ties

_____ 11. If a building is 30' long with a rafter slope of 9 on 12, the common difference in length of a 16" OC gable stud is _____*B*_____ .

    A. 9"

    B. 12"

    C. 16"

    D. 15"

_____ 12. A true gambrel has _____*D*_____ .

    A. a knee wall under the rafter intersections

    B. ceiling joists connecting middle rafter intersections

    C. rafter intersections laid out on a semicircle

    D. all of the above

Completion

Complete each sentence by inserting the best answer on the line near the number.

_____ 1. The most common style of roof used is the _Gamble_ roof.

_____ 2. On an normal gable roof, the _Common_ rafter is the only type of rafter needed to be laid out.

___Ridge___ 3. Although not absolutely necessary for structural purposes, the _____ simplifies rafter erection.

___Stock___ 4. When laying out the rafter that is to be used as a pattern, be sure to select the _____ piece possible.

___Frame Squ___ 5. The faster, more accurate method for laying out a rafter uses a _____ and the rafter tables to find rafter length.

___Fram Squ___ 6. Rafter tables come in booklet form and are also stamped on one side of a _____ .

___wall plate___ 7. On roofs with moderate slopes, the length of the level cut of the seat can be the width of the _____ .

___Horizon___ 8. The rafter projection given on a set of prints is usually given in terms of a _____ measurement.

___½ off___ 9. If a ridgeboard is used, the rafter must be _____ a distance equal to one-half the thickness of the ridgeboard.

___Rafter___ 10. All ridgeboard joints should be centered on a _____ unless a scab is used.

___gable___ 11. The first and last rafters of a gable roof are commonly called _____ rafters.

___Look out___ 12. When an overhang is required at the rakes, horizontal structural members called _____ must be installed.

___gable stub___ 13. When installing _____ , care must be taken not to force the end rafters up and create a crown in them.

___Common diffie___ 14. The _____ is helpful in determining the lengths of gable studs.

_____ 15. The best way to determine gambrel rafter lengths is to use the _____ .

Identification: Gable and Gambrel Roof Parts

Identify each term, and write the letter of the best answer on the line next to each number.

___D___ 1. ridge

___E___ 2. common rafter

___G___ 3. plumb cut or ridge cut

___A___ 4. seat cut or bird's mouth

___F___ 5. tail or overhang

___C___ 6. collar ties

___B___ 7. gable studs

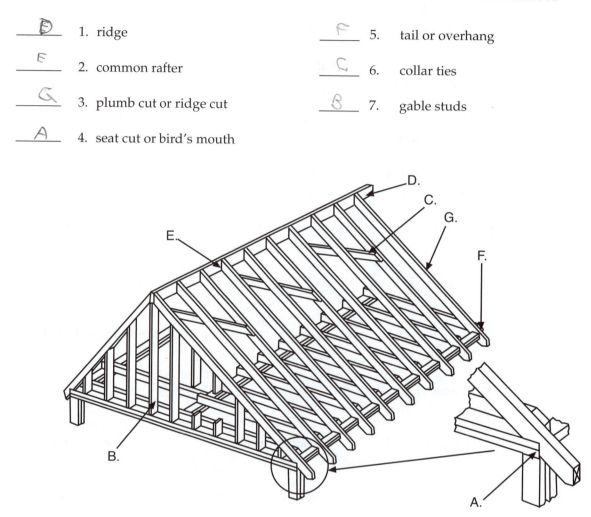

Math Problem-Solving

Solve the following math problems.

_____ 1. What is the unit length of a roof with slope of 7 on 12 to the nearest hundredth?

_____ 2. What is the run, in decimal form, of a gable roof if the width of the building is 26'-5"?

_____ 3. What is the length of a common rafter with a run of 14'-6" and a slope of 6 on 12? Answer in terms of inches to the nearest sixteenth.

_____ 4. What is the common difference in length of a gable stud spaced 16" OC fit to a rafter with a slope of 15"?

_____ 5. What is the length of the gambrel rafter R shown in the figure? Answer in terms of inches to the nearest sixteenth.

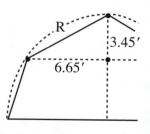

43 Hip Roofs

Multiple Choice

Write the letter for the best answer on the line next to the number of the sentence.

_____ 1. Hip roofs are _____ .
   A. easier to frame than gable roofs
   B. more complicated than gable roofs to frame
   C. the most common style of roof used
   D. sloped in one direction only

_____ 2. Hip rafters are required _____ .
   A. where the slopes of a hip roof meet
   B. on all roof framing
   C. on saltbox style roofs
   D. at right angles from the plates to the common rafters

_____ 3. In comparison to the common rafters on the same roof, _____ .
   A. hip rafters must rise the same but with fewer steps
   B. hip rafters have a decreased unit of run
   C. the slope of a hip rafter is much steeper
   D. the unit of run of the hip rafter is increased

_____ 4. If the pitch of a hip roof is a 6″ rise per unit of run, the hip rafter would be laid out by holding the square at _____ .
   A. 6 and 12
   B. 6 and 14
   C. 6 and 17
   D. 6 and 24

_____ 5. The ridge cut of a hip rafter _____ .
   A. is a compound angle
   B. is called the cheek cut
   C. may be called a side cut
   D. all of the above

_____ 6. When calculating the length of a hip rafter using the tables on the framing square, _____.
   A. the numbers found on the square are divided by the overall run of the rafter
   B. the numbers from the second line are used
   C. the common difference must first be determined
   D. all of the above

_____ 7. When laying out the seat cut of a hip rafter,_____ .
   A. consideration must be given to fitting it around the corner of the wall
   B. consideration must be given to dropping the hip rafter
   C. the next to the last plumb line laid out is used as the plumb cut for the seat of the rafter
   D. the numbers from the third line of the framing square are used

_____ 8. When the top-edge corners of a hip rafter are beveled flush with the roof, it is called
_____ .
   A. raising the hip
   B. dropping the hip
   C. backing the hip
   D. opposing the hip

_____ 9. A double-cheek cut is usually made at the _____ .
   A. tail cut of a hip rafter
   B. seat cut of a hip rafter
   C. seat cut of a hip-jack rafter
   D. tail cut of a hip-jack rafter

_____ 10. The hip-jack rafter _____ .
   A. runs at right angles from the hip rafter to the plate
   B. has the same unit of run as the hip rafter
   C. is actually a shortened common rafter
   D. has a double-cheek cut

_____ 11. The run of hip-jack rafters meet the run of the hip rafter at _____ .
   A. 22½°
   B. 45°
   C. 60°
   D. 90°

_____ 12. The line length of a hip roof ridge is _____ .
   A. the same as the overall length of the building
   B. found by subtracting one-half the width of the building from its length
   C. the length of the building minus its width
   D. determined from information on the rafter tables of the framing square

Identification: Hip Roof Parts

Identify each term, and write the letter of the best answer on the line next to each number.

_____ 1. hip jack rafter   _____ 4. hip rafter

_____ 2. ridge   _____ 5. plate

_____ 3. common rafter

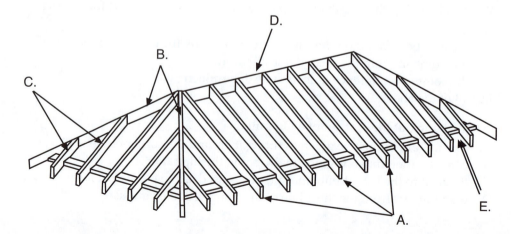

Math Problem-Solving

Solve the following math problems using the information found in the figure shown.

_____ 1. What is the unit length of the common rafter? Answer in decimal form to the nearest thousandth.

_____ 2. What is the unit length of the hip rafter? Answer in decimal form to the nearest thousandth.

_____ 3. Find the length of the common rafter C. Answer in inches to the nearest sixteenth.

_____ 4. Find the length of the hip rafter H. Answer in inches to the nearest sixteenth.

_____ 5. Find the length of the jack rafter J. Answer in inches to the nearest sixteenth.

_____ 6. Find the common difference in length that L is shorter than J. Answer in inches to the nearest sixteenth.

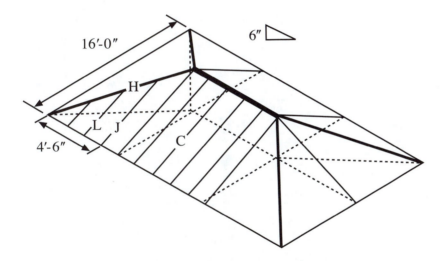

Discussion

Write your answer(s) on the lines below.

1. When planning a home design, what factors might go into the decision of choosing a hip roof?

_____

_____

_____

44 The Intersecting Roof

Multiple Choice

Write the letter for the best answer on the line next to the number of the sentence.

_____   1. Intersecting roofs usually have _____ .
   A. hip rafters
   B. valley rafters
   C. purlins
   D. kneewalls

_____   2. If the heights of two intersecting roofs are different, _____ .
   A. supporting and shortened valley rafters must be used
   B. a valley cripple-jack rafter must be used
   C. valley-jack rafters must be used
   D. all of the above

_____   3. The unit of run for valley rafters is _____ .
   A. 12
   B. the same as common rafters
   C. called the minor span
   D. the same as the hip rafter

_____   4. In order for the valley rafter to clear the inside corner of the wall, _____ .
   A. it must be dropped like the hip rafter
   B. the seat cut must be extended
   C. the seat cut must be raised
   D. the wall plate is notched

_____   5. The supporting valley rafter is shortened _____ .
   A. by ½ the thickness of the ridge board
   B. by the thickness of the ridge board
   C. by ½ the 45° thickness of the ridge board
   D. only if a shortened valley rafter is needed

_____   6. The total run of the supporting valley rafter is _____ .
   A. the run of the common rafter of the main roof
   B. the run of the common rafter of the smaller roof
   C. known as the major span
   D. A and C

_____   7. The tail cut of the valley rafter is _____ .
   A. identical to that of a hip rafter
   B. a single cheek cut
   C. a square cut
   D. a double cheek cut that angles inward

_____ 8. The unit of run used to determine the length of the valley-jack rafter is _____.
    A. the same one used for valley rafters
    B. the same one used for hip rafters
    C. the same one used for common rafters
    D. 17″ if the step-off method is used

_____ 9. The total run of any valley-jack rafter is _____.
    A. the same as the common rafters
    B. known as the minor span
    C. that of the common rafter minus the horizontal distance it is located from the corner of the building
    D. that of the supporting valley rafter minus the horizontal distance it is located from the end of the building

_____ 10. Hip valley cripple-jack rafters cut between the same hip and valley rafters _____.
    A. are the same length
    B. have square cheek cuts
    C. differ in size by the same common difference found on the rafter table
    D. A and B

Identification: Intersecting Roof Parts

Identify each term, and write the letter of the best answer on the line next to each number.

_____ 1. ridge of major span   _____ 6. hip valley cripple-jack rafter

_____ 2. supporting valley rafter   _____ 7. hip-jack rafter

_____ 3. shortened valley rafter   _____ 8. valley-jack rafter

_____ 4. common rafter   _____ 9. valley cripple-jack rafter

_____ 5. ridge of minor span   _____ 10. hip rafter

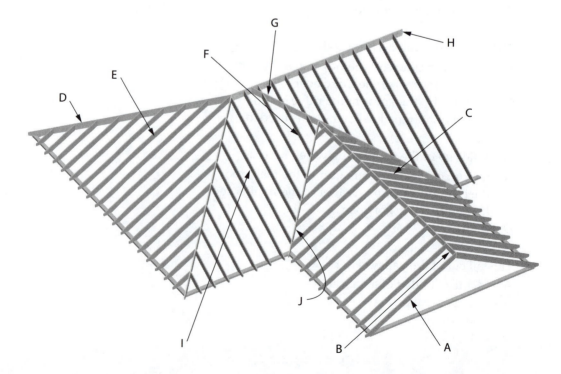

Math Problem-Solving

Solve the following math problems using the information found in the figure shown. All answers in terms of inches to the nearest sixteenth.

_____ 1. Find the length of supporting valley rafter V.

_____ 2. Find the length of shortened valley rafter W.

_____ 3. Find the length of valley jack rafter 1.

_____ 4. Find the length of valley jack rafter 2.

_____ 5. Find the length of gable stud 3.

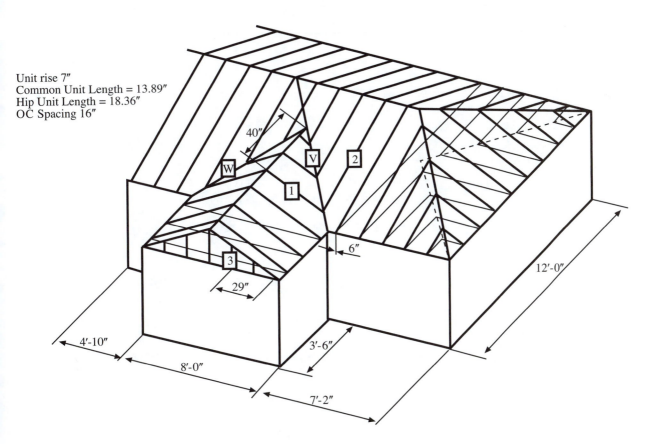

Unit rise 7"
Common Unit Length = 13.89"
Hip Unit Length = 18.36"
OC Spacing 16"

Discussion

Write your answer(s) on the lines below.

1. List four organizing thoughts that will help eliminate confusion concerning layout of so many different kinds of rafters.

   a. _____

   _____

   b. _____

   _____

   c. _____

   _____

   d. _____

45 Shed Roofs, Dormers, and Other Roof Framing

Multiple Choice

Write the letter for the best answer on the line next to the number of the sentence.

_____ 1. The purpose of a dormer is to provide _____ .
   A. light
   B. ventilation
   C. architectural appeal
   D. all of the above

_____ 2. The rafter that may have two bird's mouths is the _____ .
   A. shed dormer rafter
   B. shed roof rafter
   C. hip-valley cripple jack
   D. all of the above

_____ 3. The layout of a shed dormer rafter is most similar to a _____ .
   A. hip rafter
   B. common rafter
   C. hip jack rafter
   D. collar tie

_____ 4. The shed dormer member that requires a compound angle to be cut is the _____ .
   A. ridgeboard
   B. lower end of a rafter fit against an existing roof
   C. upper end of a rafter fit against an existing roof
   D. valley jack

_____ 5. The shed dormer member that requires a plumb cut that is laid out using two setups of the framing square is the _____ .
   A. ridgeboard
   B. lower end of a rafter fit against an existing roof
   C. upper end of a rafter fit against an existing roof
   D. valley jack

Completion

Complete each sentence by inserting the best answer on the line near the number.

_____ 1. The shed roof slopes in only _____ direction.

_____ 2. The unit of run for a shed roof is _____ inches.

_____ 3. A _____ is a framed projection above the plane of the roof containing one or more windows.

_____ 4. The length of a shed roof rafter can be determined by using the rafter tables for a _____ rafter.

_____ 5. Shed roofs are framed by _____ the rafters into the plate at the designated spacing.

_____ 6. When framing a shed roof, it is important that the plumb cut of the seat be kept snug against the _____.

_____ 7. The rafters on both sides of a dormer opening must be _____.

_____ 8. Top and bottom _____ of sufficient strength must be installed when dormers are framed with their front wall partway up the main roof.

_____ 9. In most cases, shed dormer roofs extend to the ridge of the main roof in order to gain enough _____.

Math Problem-Solving

Solve the following math problems using the information found in the figure shown. All framing is with dimension lumber 16″ OC.

_____ 1. What is the run of the hip rafter?

_____ 2. What is the rough length of the ridgeboard R?

_____ 3. How much must be added to the end of the ridgeboard near A?

_____ 4. How much must be added to the end of the ridgeboard near B?

_____ 5. What is the actual length of ridgeboard R? Answer in terms of feet-inches to the nearest sixteenth.

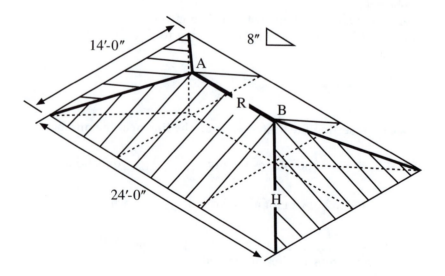

Discussion

Write your answer(s) on the lines below.

1. What is a disadvantage of building an intersecting roof after the main roof has been framed and sheathed?

_____

_____

_____

46 Trussed Roofs

Multiple Choice

Write the letter for the best answer on the line next to the number of the sentence.

_____ 1. Truss use in roof framing _____ .
   A. increases the home's usable attic space
   B. slows down the home's construction
   C. eliminates the need for load-bearing interior partitions below
   D. all of the above

_____ 2. The diagonal parts of roof trusses are called _____ .
   A. web members
   B. knee supports
   C. stringers
   D. sleepers

_____ 3. The upper cords of a roof truss act as _____ .
   A. ceiling joists
   B. rafters
   C. collar ties
   D. a ridge board

_____ 4. Most trusses used today _____ .
   A. are designed by the carpenter
   B. are made in fabricating plants
   C. are built on the job
   D. do not contain gusset plates

_____ 5. Approved designs and instructions for job-built trusses are available from the
   _____ .
   A. American Plywood Association
   B. Truss Plate Institute
   C. American Hardwood Association
   D. A and B

_____ 6. The most common truss design is the _____ truss.
   A. howe
   B. pratt
   C. fink or W
   D. scissors

_____ 7. Carpenters are more involved in the _____ of trusses.
   A. erection
   B. design
   C. construction
   D. fabrication

8. Failure to properly erect and brace a trussed roof could result in _____ .
   A. collapse of the structure
   B. loss of life or serious injury
   C. loss of time and material
   D. all of the above

9. The truss bracing system depends a great deal on _____ .
   A. the use of 1″ × 4″ temporary bracing
   B. scabs nailed to the end of the building
   C. how well the first truss is braced
   D. 8d duplex finish nails

10. As bracing is installed, it is important that _____ .
   A. exact spacing be maintained
   B. spacing be continually readjusted
   C. it is only applied to two planes of the truss assembly
   D. it has a maximum length of no more than 8′

Completion

Complete each sentence by inserting the best answer on the line near the number.

1. The plane of the top cord is known as the _____ plane.

2. The plane of the bottom cord is known as the _____ plane.

3. It is recommended that continuous lateral bracing be placed within _____ inches of the peak on the top cord plane.

4. Diagonal bracing between the rows of lateral bracing should be placed on the _____ of the top plane.

5. Continuous lateral bracing of the bottom cord must be applied to maintain the proper _____ .

6. Bottom cord bracing is nailed to the _____ of the bottom cord.

7. Bottom cord bracing should be installed at intervals no greater than _____ to _____ feet along the width of the building.

8. Rated panels of plywood and _____ are commonly used to sheath roofs.

9. In post and beam construction where roof supports are spaced further apart, _____ is used for roof sheathing.

10. Panel clips, tongue and groove edges, or other adequate blocking must be used when _____ exceed the indicated value of the plywood roof sheathing.

Math Problem-Solving

Solve the following math problems using the following information:

Rectangular building 28′ × 42′, ridgeboard length 44′-6″, common rafter length 187⅞″

_____ 1. Neglecting the rake over hang rafters, how many gable rafters are needed?

_____ 2. If this building were to be a hip roof, how many common rafters would be needed?

_____ 3. How many rated panel sheets to sheath the gable roof?

_____ 4. How many rated panel sheets to sheath the hip roof?

_____ 5. How many trusses would be used if they are placed 24″ OC?

Discussion

Write your answer(s) on the lines below.

1. List the advantages of using trusses over conventional roof framing.

_____

_____

_____

2. What are disadvantages to using roof trusses? How might this affect future remodeling or sale of the home?

_____

_____

_____

47 Stairways and Stair Design

Multiple Choice

Write the letter for the best answer on the line next to the number of the sentence.

_____ 1. The tread run is _____ .
   A. the part of the tread that extends beyond the riser
   B. the horizontal distance between the faces of the risers
   C. the finish material that covers the vertical distance from one step to another
   D. a nonskid material applied to tread

_____ 2. The housed finished stringer is _____ .
   A. usually fabricated in a shop
   B. usually built on the job site
   C. always used on service stairs
   D. installed by the framing crew

_____ 3. Dadoes routed into the sides of the housed finished stringer _____ .
   A. reduce the stair's rise
   B. increase the headroom
   C. support and house the risers and treads
   D. all of the above

_____ 4. The preferred angle for ease of stair climbing is between _____ degrees.
   A. 20 and 25
   B. 25 and 30
   C. 30 and 35
   D. 35 and 40

_____ 5. To determine the unit rise, the _____ must be known.
   A. total run
   B. total rise
   C. riser thickness
   D. tread thickness

_____ 6. To determine riser height without a calculator, _____ used.
   A. a sliding T bevel and a level are
   B. the common rafter table on the framing square is
   C. a story pole and a set of dividers are
   D. a plumb bob and a line level are

_____ 7. The sum of one rise and one tread should equal _____ .
   A. between 17 and 18
   B. between 18 and 19
   C. between 19 and 20
   D. less than 17

_____ 8. Decreasing riser height _____.
   A. decreases the run of the stairs
   B. increases the run of the stairs
   C. uses less space
   D. makes the stairs more difficult to climb

_____ 9. Increasing stair riser height typically _____.
   A. decreases the run of the stairs
   B. makes the stairs more difficult to climb
   C. decreases the overall space occupied by the stairs
   D. all of the above

_____ 10. Most building codes call for a minimum of _____ of headroom for stairways.
   A. 6'-2"
   B. 6'-4"
   C. 6'-6"
   D. 6'-8"

Completion

Complete each sentence by inserting the best answer on the line near the number.

_____ 1. The stairwell is framed at the same time as the _____.

_____ 2. _____ stairs extend from one habitable level of the house to another.

_____ 3. _____ stairs extend from a habitable level to a nonhabitable level.

_____ 4. In residential construction, stairways should not be less than _____ inch(es) wide.

_____ 5. A _____ stairway is continuous from one floor to another without any turns or landings.

_____ 6. Stairs that have intermediate landings between floors are called _____ stairs.

_____ 7. An L-type platform stair changes direction _____ degrees.

_____ 8. A platform stairway that changes directions 180° is called a _____ -type stairway.

_____ 9. A _____ staircase gradually changes directions as it ascends from one floor to another.

_____ 10. A _____ stairway is constructed between walls.

_____ 11. _____ stairways have one or both sides exposed to a room.

_____ 12. The vertical distance between finished floors is the total _____ of the stairway.

_____ 13. The total horizontal distance that the stairway covers is known as the total _____ .

_____ 14. The vertical distance from one step to another is the _____ .

_____ 15. The opening in the floor that the stairway passes through is the _____ .

Math Problem-Solving

Solve the following math problems for stairs with a total rise of 8'-4".

_____ 1. What is the recommended starting dimension for unit rise when performing stair calculations?

_____ 2. What is the minimum number of risers in stairs?

_____ 3. What would be the largest unit rise allowed to the nearest sixteenth?

_____ 4. Using the 24-25 method, what would be the smallest unit run (tread)?

_____ 5. What is the total run of the stairs, if the unit run from question 4 were used? Assume the upper construction is used as a riser.

48 Stair Layout and Construction

Multiple Choice

Write the letter for the best answer on the line next to the number of the sentence.

_____ 1. When laying out stair carriages, make sure _____ .
   A. riser heights are greater than tread widths
   B. all riser heights are the same
   C. all tread widths are the same
   D. B and C

_____ 2. When scaling across a framing square to determine the rough length of a stair carriage, use the side of the square that is graduated in _____ of an inch.
   A. sixteenths
   B. eighths
   C. tenths
   D. twelfths

_____ 3. When stepping off the stair carriage with a framing square, _____ .
   A. use stair gauges
   B. lay out the rise with the square's tongue
   C. lay out the tread run with the square's blade
   D. all of the above

_____ 4. To be sure the bottom riser is the same height as all the other risers it may be necessary to _____ .
   A. add blocking under the stair carriage bottom
   B. cut a certain amount off the stair carriage bottom
   C. alter the thickness of the first tread
   D. alter the thickness of all treads but the first

_____ 5. The top riser is equalized by _____ .
   A. lowering and cutting the level line at the top of the stair carriage
   B. altering the top treads thickness
   C. fastening the stair carriage at the proper height in relation to tread and finish floor thickness
   D. adding blocking to the top of the stair carriage

_____ 6. Residential staircases of average width usually_____ .
   A. require three carriages
   B. only need two carriages
   C. are less than 36″ wide
   D. are more than 48″ wide

_____ 7. Built-up stair carriages are used to _____ .
   A. simplify drywall application
   B. conserve wood
   C. avoid dropping the stair carriage
   D. increase the stairs' strength

_____ 8. If drywall is applied after the stairs are framed, _____ .
   A. repair or remodeling work becomes more difficult than if it were applied before framing
   B. time is saved
   C. blocking is required between studs in back of the stair carriage
   D. no blocking is needed

_____ 9. U-type stairways usually have the landing _____ of the stairs.
   A. near the bottom
   B. in the middle
   C. near the top
   D. A and C

_____ 10. According to most codes, any flight of stairs that has a vertical distance of 12' or more must have _____ .
   A. double railings on each side
   B. a width of at least 4'
   C. at least one landing
   D. a tread run of at least 12"

Math Problem-Solving

Solve the following math problems using the following information: total stair rise is 103", headroom needs to be 82", and the upper construction is 13½".

_____ 1. What would be the minimum number of risers recommended for the stairs?

_____ 2. What would be the unit rise of the stairs?

_____ 3. Using the 17-18 method, what would be the maximum tread width?

_____ 4. What is the sum of headroom and upper construction?

_____ 5. What is the minimum stairwell length necessary to accommodate the headroom? The top riser is part of the upper construction.

Discussion

Write your answer(s) on the lines below.

1. What important safety rules must be considered when laying out stairs?

   _____

   _____

   _____

2. Although winding stairs are not recommended, what are some of the rules that must be followed if they are used?

   _____

   _____

   _____

49 Thermal and Acoustical Insulation

Multiple Choice

Write the letter for the best answer on the line next to the number of the sentence.

_____ 1. Organic insulation materials are treated to make them resistant to _____ .
      A. fire
      B. insects
      C. vermin
      D. all of the above

_____ 2. _____ insulation would be a good choice for insulating the sidewalls of an older, uninsulated home.
      A. Loose-fill
      B. Rigid
      C. Flexible
      D. Reflective

_____ 3. Rigid insulation is usually made of _____ .
      A. vermiculite
      B. glass wool
      C. fiber or foamed plastic material
      D. rock wool

_____ 4. Reflective insulation should be installed _____ .
      A. tight against the siding
      B. facing an air space with a depth of ¾″ or more
      C. in contact with the foundation wall when used in basements or crawl spaces
      D. only in homes without air conditioning

_____ 5. Foamed-in-place insulation _____ .
      A. expands on contact with the surface
      B. is made from organic fibers
      C. contains glass wool
      D. is commonly used for sheathing and decorative purposes

_____ 6. To reduce heat loss, all _____ that separate heated from unheated areas must be insulated.
      A. walls
      B. ceilings
      C. roofs and floors
      D. all of the above

_____ 7. When installing flexible insulation between floor joists over crawl spaces,
_____ .

    A. the vapor barrier faces the heated area
    B. the vapor barrier faces the ground
    C. remove the vapor barrier
    D. at least a ¾″ air space must be maintained between the joists and the insulation

_____ 8. The resistance to the passage of sound through a building section is rated by its
_____ .

    A. Sound Transmission Class
    B. Impact Noise Rating
    C. Sound Absorption Class
    D. Impact Transmission Class

Completion

Complete each sentence by inserting the best answer on the line near the number.

_____ 1. _____ insulation prevents the loss of heat in cold seasons and resists the passage of heat into air-conditioned areas in the hot seasons.

_____ 2. _____ insulation reduces the passage of sound from one area to another.

_____ 3. Proper ventilation must be provided within the building to remove _____ that forms in the space between the cold surface and the thermal insulation.

_____ 4. If confined to small spaces in which it is still, _____ is an excellent insulator.

_____ 5. Aluminum foil is used as an insulating material that works by _____ heat.

_____ 6. The higher the R-value number of insulation, the more _____ the material is.

_____ 7. The use of _____ studs in exterior walls permits the installation of R-19 insulation.

_____ 8. _____ insulation is manufactured in blanket and batt form.

_____ 9. The _____ on both sides of blanket insulation facing are used for fastening it to studs or joists.

_____ 10. Batt insulation is available in thicknesses of up to _____ inches.

Math Problem-Solving

Solve the following math problems using the following information: rectangular 32′ × 60′ building with 8′ walls.

_____ 1. How many SF (square feet) of insulation is needed for the exterior walls? Deduct 6-15 SF windows, 3-12 SF windows, 2-42 SF windows and 2-20 SF doors.

_____ 2. If each bundle of batt insulation covers 155 SF, how many bundles would be needed?

_____ 3. How many sheets of high-R sheathing are needed to cover the exterior walls? Include 1′ extra height to cover the band joist area.

_____ 4. How many SF of insulation is needed to cover the ceiling?

_____ 5. If blown-in insulation costs $0.75 per square foot, how much will it cost to insulate the ceiling?

50 Condensation and Ventilation

Multiple Choice

Write the letter for the best answer on the line next to the number of the sentence.

_____ 1. When the temperature of moisture-laden air drops below its dew point, _____ occurs.
   A. evaporation
   B. condensation
   C. vaporization
   D. dehydration

_____ 2. Condensation of water vapor in walls, attics, roofs, and floors _____ .
   A. increases the R-value of any insulation it comes in contact with
   B. prevents the wood from rotting
   C. leads to serious problems
   D. only occurs in warmer climates

_____ 3. To prevent the condensation of moisture in a building, _____ .
   A. reduce the moisture in the warm, inside air
   B. ventilate the attic
   C. install a barrier to the passage of water vapor
   D. all of the above

_____ 4. In a well-ventilated area, _____ .
   A. condensed moisture is removed by evaporation
   B. moisture is forced back into the insulation
   C. insulation is not necessary
   D. condensation is at its worst

_____ 5. Virtually all moisture migration into insulation layers can be eliminated by _____ .
   A. airtight construction techniques
   B. not using a vapor retarder on the ceiling
   C. proper ventilation
   D. all of the above

_____ 6. On roofs where the ceiling finish is attached to the rafters and insulation is installed, _____ .
   A. an air space is not needed
   B. a moisture barrier should be placed on the cold side of the insulation
   C. a well-vented air space of at least 1" is necessary between the insulation and the roof sheathing
   D. no ventilation is needed

_____ 7. The most effective method of ventilating an attic is to install _____.
   A. continuous soffit vents
   B. continuous ridge vents
   C. larger triangular gable end louvers
   D. A and B

_____ 8. Generally speaking, as the temperature of the air decreases the relative humidity _____.
   A. decreases
   B. increases
   C. remains unchanged
   D. is improved

_____ 9. The use of a ground cover in crawl spaces _____.
   A. is not recommended
   B. contributes to the decay of framing members
   C. eliminates the need for ventilators
   D. allows for the use of a smaller number of ventilators

_____ 10. The minimum free-air area for attic ventilators is based on the _____.
   A. total square footage of the home
   B. ceiling area of the rooms below
   C. attic floor area
   D. total roof surface

Completion

Complete each sentence by inserting the best answer on the line near the number.

_____ 1. When air temperature reaches _____, condensation begins.

_____ 2. The boundary between heated and unheated spaces is referred to as _____.

_____ 3. The use of _____ prevents moisture from entering areas where condensation can occur.

_____ 4. The most commonly used material for a vapor barrier is _____.

_____ 5. _____ allow for ventilation of cavities created by hip-jack rafters.

_____ 6. _____ the seams of the exterior wall sheathing is an effective method of airtight construction.

_____ 7. The minimum free-air area of attic ventilators needed for a home with a ceiling area of 1,500 square feet is _____ square feet.

_____ 8. _____ cause water to back up the roof and into the building.

_____ 9. _____ prevent blown-in insulation from entering the soffit at the eaves.

_____ 10. _____ built into a building allows moisture that finds its way into a building to evaporate at some later time.

Math Problem-Solving

Solve the following math problems.

_____ 1. What is the sum of 50 and –15?

_____ 2. What is the difference in temperature between 60° and –5°?

_____ 3. What is the average temperature of 60° and 10°?

_____ 4. What is the average daily temperature if the daily high is 60° and the low is –10°?

_____ 5. A *degree day* is the difference between 65° and the average daily temperature. If the daily high temperature is 50° and daily low temperature is 32°, what is the degree day for this day?

Discussion

Write your answer(s) on the lines below.

1. List the necessary steps that must be done during the construction process to eliminate condensation problems in a home.

_____

_____

_____

Section 2: Building for Success

COMMUNICATION SKILLS IN CONSTRUCTION

The world of construction has always relied on strong oral and written communication skills to relay information. Today we communicate electronically more than ever before. Yet the most effective exchange of information is still based on good speaking, writing, and listening skills.

Along with the personal computer, fax machines, fiber optics, and telecommunications, there is still a need to develop strong oral and written communication skills. These skills form the basis for using high-technology achievements successfully.

Oral communication has always been important in providing the desired "personal touch" for customers. The vast majority of construction projects require good oral and written interaction both in the office and at the job site. As we become a more service-oriented nation, we must continue to value the power of the spoken and written word in conducting business, To enhance the success of the project, the builder, sub-contractor, or technician on the job must maintain exactness and clarity when dealing with clients.

The diverse populations that make up the nation's workforce today bring new dimensions to effective communication. Construction personnel represent more cultures, education levels, and diversities than ever before. This alone calls for a greater need to promote clear and concise oral and written communication at the workplace. These new dimensions call for increased abilities in reading, writing, and listening.

Technicians today are more involved in the planning, coordinating, and marketing as well as the construction of a project. This will continue to require more verbal and writing skills. Legalities, warranties, quality assurances, and guarantees have taken on expanded roles. Interpreting codes, specifications, statutes, and other stipulations demands better reading and listening skills. Mistakes due to errors in reading or interpretation can be costly and even endanger lives. All workers must develop a proficiency in effective communication.

One of the biggest challenges to young people entering the industry today appears to be in the area of using appropriate language. The use of inappropriate or abusive language in the business world can seriously restrict the success of the project. Associated with this problem is the concern for forming appropriate business attitudes. To illustrate this concern, we need to look at the issue from the client's viewpoint.

Let's take, for example, the family that is having a new house built in a community. Excluding a business venture, this will be the largest investment they make in their lifetime. They are envisioning a new abode where they will be rearing their children, sharing family experiences, and possibly spending their retirement years. They take it seriously. Beginning with the planning phase, their expectation level is high.

Now enter the builders and workers representing the various trades. These representatives from the construction industry have the responsibility of being professional. They should have the trust of the customer as part of the construction package. An agreement has been reached. The builder and the team are obligated to provide the best quality for the price being paid. This must be communicated.

The builder must demonstrate professional attitudes and use professional language. If the builder or construction representative uses abusive or profane language on the job site, it sends a negative message to the client. The message will say, "This project is not important; it's just another routine job." That, in turn, will probably tell the client that the quality is less than what should be expected.

Workers with the proper attitude and command of the language will not spit chewing tobacco on the subfloor or present themselves vocally as crude or uncaring. We must remember that clients will advertise for the builder whether the builder likes it or not. They will share with others how they perceived the builder's professionalism and work ethics in addition to the quality of work. Not all communication is conducted orally or in writing. We also communicate who we are by our performance. People read each other continually.

We must ask ourselves, "How can I improve my communication skills?" All people should take personal inventories to see how well equipped they are. Some questions and suggestions follow that may help students determine their own strengths or weaknesses in personal communication skills.

- Can I list, in writing, my career goals?
- How much do these goals involve oral and written communications?
- What is my current skill level in communicating?
- Do I listen, write, and speak well? Is my vocabulary acceptable?
- What does my past communication performance in school indicate?
- Do I understand what personal communication skills will be demanded of me in a construction career?
- Do I understand how communication skills will describe my professional image, either positively or negatively?
- Do I know someone I can talk to that knows what communication skills will help me reach my goals?
- Do I know how these skills can be obtained?
- Once I understand what good communication skills will be required of me to be successful, will I desire to obtain those skills?

An important recommendation to any student pursuing a career in construction is to talk to knowledgeable people who can provide accurate information about the necessary communication requirements for the desired job. Trade representatives, counselors, parents, teachers, and the CEOs of construction businesses will be the people with that knowledge.

Someone once said, "It is not so much what we know as how well we use what we know." A personal vocabulary does not have to equal the volume of a dictionary. But words must be selected carefully. A good policy may be one that says, "If you would not write it and sign it, do not say it."

In conclusion, the element of effective communication must never be diminished or disregarded. It is a vital part of good business ethics. it has become even more important as we strive for success in the workplace.

Focus Questions:
For individual or group discussion

1. Address each question presented in the preceding study.
2. What would be the consequences of not developing good communication skills as you pursue a career in construction?

SECTION THREE
EXTERIOR FINISH

51 Asphalt Shingles

Multiple Choice

Write the letter for the best answer on the line next to the number of the sentence.

_____ 1. When installing metal drip edge, _____ .
   A. tightly butt the end joints
   B. only use 1¼" roofing nails
   C. the roofing nails must be the same material as the drip edge
   D. space the nails every 12"

_____ 2. Asphalt shingle underlayment should _____ .
   A. be an entirely moisture-proof membrane
   B. allow the passage of water vapor
   C. be applied in vertical rows
   D. usually be as heavy a weight of felt as is available

_____ 3. After the application of underlayment, it is recommended that _____ .
   A. metal drip edge be applied to the rakes
   B. metal drip edge be applied to the eaves
   C. the ridge vent be installed
   D. all of the above

_____ 4. Organic shingles have a base made of _____ .
   A. glass fibers
   B. heavy asphalt-saturated paper felt
   C. shredded thatching straw
   D. rubber

_____ 5. Mineral granules are used on the surface of asphalt shingles to _____ .
   A. provide weatherproofing qualities
   B. provide a good surface for asphalt cement to adhere to
   C. reduce its weight
   D. protect the shingle from the sun

_____ 6. _____ is generally determined by weight per square.
   A. Fire resistance
   B. Shingle quality
   C. Self-sealing ability
   D. Coverage

_____ 7. On long roofs, accurate vertical alignment is ensured by _____ .
   A. starting from either rake
   B. working right to left
   C. working left to right
   D. starting at the center and working both ways

_____ 8. It is recommended that no rake tab be less than _____ in width.
  A. 2″
  B. 3″
  C. 4″
  D. 5″

_____ 9. The purpose of the starter course is to _____ .
  A. back up and fill in the spaces between the tabs on the first row of shingles
  B. eliminate the need of installing a drip edge
  C. reduce frost build up underneath the shingles
  D. all of the above

_____ 10. When fastening asphalt shingles, it is important to _____ .
  A. use a minimum of three fasteners in each strip shingle
  B. follow the manufacturer's recommendations for application
  C. use a fastener long enough to penetrate the sheathing at least ⅜″
  D. not use power nailers

_____ 11. The most commonly used asphalt shingles have a maximum exposure of _____ inches.
  A. 3
  B. 5
  C. 7
  D. 9

_____ 12. When snapping a long chalk line, many times it is necessary to _____ .
  A. first wet the line
  B. snap the line from the side closest to the chalk box
  C. hold the line against the roof with your thumb at about center and strike of the line
  D. simultaneously strike both sides of the line

_____ 13. When installing the ridge caps _____ .
  A. coat the last two fasteners with asphalt cement
  B. start the shingles on the end away from the prevailing winds
  C. it may be necessary to warm them in cold weather
  D. all of the above

_____ 14. The maximum roof angle recommended for normal asphalt shingle application is _____ degrees.
  A. 45
  B. 50
  C. 55
  D. 60

_____ 15. Nails driven into asphalt shingles should be _____ the asphalt sealing strip.
  A. just above
  B. on
  C. just below
  D. anywhere near

Matching

Write the letter for the best answer on the line near the number to which it corresponds.

_____ 1. square

_____ 2. electrolysis

_____ 3. end lap

_____ 4. deck

_____ 5. course

_____ 6. flashing

_____ 7. asphalt cements

_____ 8. top or head lap

_____ 9. exposure

_____ 10. underlayment

A. material used to protect a roof before shingles are applied

B. strips of thin sheet metal used to make watertight joints

C. trowel applied adhesives used to bond asphalt roofing products

D. the amount of roofing required to cover 100 square feet

E. a reaction that occurs between unlike metals when wet

F. horizontal distance the ends of roofing in the same course overlap

G. the amount of roofing in each course subjected to the weather

H. the wood roof surface to which roofing is applied

I. shingle height minus the exposure

J. a horizontal row of shingles

Math Problem-Solving

Solve the following math problems. Round each answer to the nearest whole number.

_____ 1. What is the area of a triangle with a height of 9'-7" and a base of 24'-5"?

_____ 2. What is the actual area of a gable roof with a rafter length of 15'-6" and ridgeboard length of 40'-6"?

_____ 3. For purposes of estimating roof sheathing, what is the area of a gable roof with a rafter length of 15'-6" and ridgeboard length of 40'-6"?

_____ 4. If the roof in question 3 were changed to a hip roof, estimate the number of rated panels needed.

_____ 5. What is the line length of the entire rafter if its slope is 5 on 12, run is 12'-6", and projection is 12"?

52 Roll and Tile Roofing

Multiple Choice

Write the letter for the best answer on the line next to the number of the sentence.

_____ 1. The number of squares in a roll of double coverage roll roofing is _____ .
   A. ¼
   B. ¼
   C. ¾
   D. 1

_____ 2. The lowest recommended temperature for installing roll roofing is _____ degrees.
   A. 0°C
   B. 32°F
   C. 45°F
   D. 50°F

_____ 3. The minimum end lap of roll roofing is _____ .
   A. 6″
   B. 9″
   C. 12″
   D. 18″

_____ 4. Cement roof tiles are installed with _____ .
   A. masonry nails
   B. roofing nails
   C. adhesive
   D. screws

_____ 5. In areas of severe wind blown rain, cement roof tiles are installed on _____ layers of underlayment.
   A. one
   B. two
   C. three
   D. four

Completion

Complete each sentence by inserting the best answer on the line near the number.

_____ 1. Roll roofing can be installed on roofs that slope as little as _____ inch of rise per foot of run.

_____ 2. A concealed nail type of rolled roofing called _____ has a top lap of 19″.

_____ 3. All kinds of rolled roofing come in rolls that are _____ inches wide.

_____ 4. Rolled roofing's coat can crack if it is applied at temperatures below _____ degrees Fahrenheit.

_____ 5. Use only the lap or quick setting cement recommended by the _____ .

_____ 6. Cement roof tiles have a life expectancy of _____ .

_____ 7. The weight of cement roof tiles is _____ than asphalt shingles.

_____ 8. When cement roof tiles are used, truss manufacturers require the trusses be _____ before interior finishes are applied.

_____ 9. Most areas use at least one layer of _____ underlayment.

_____ 10. Cement roof tiles are fastened with _____ .

Math Problem-Solving

Solve the following math problems.

_____ 1. How many squares of shingles should be estimated to cover a roof area of 1230 square feet?

_____ 2. How many squares of shingles should be estimated for a gable roof where the rafter is 18' and the ridge is 36'?

_____ 3. If there are three bundles of asphalt shingles per square, how many bundles should be estimated for a gable roof with a ridge of 42' and a rafter of 16'?

_____ 4. How many rolls of 15# roofing felt would be needed to cover a roof area of 3450 square feet?

_____ 5. How many rolls of double coverage roll roofing would be needed to cover 2560 square feet?

Discussion

Write your answer(s) on the lines below.

1. Due to the fact that rolled roofing is made by several different manufacturers, the builder must be aware of what things when applying it?

_____

_____

_____

53 Wood Shingles and Shakes

Multiple Choice

Write the letter for the best answer on the line next to the number of the sentence.

_____ 1. Most wood shingles and shakes are produced from _____ .
   A. white pine
   B. poplar
   C. western red cedar
   D. redwood

_____ 2. Wood shingles have a _____ surface.
   A. somewhat rough
   B. relatively smooth sawn
   C. highly textured, natural grain, split
   D. very smooth

_____ 3. There are _____ standard grades of wood shingles.
   A. 2
   B. 3
   C. 4
   D. 5

_____ 4. The area covered by one square of shingles or shakes depends on the _____ .
   A. amount of shingle or shake exposed to the weather
   B. weight of the square
   C. choice of underlayment
   D. choice of deck material

_____ 5. Shingles and shakes may be applied _____ .
   A. over spaced or solid roof sheathing
   B. on roofs with slopes under 3″ of rise per foot
   C. only if extra heavy underlayment is used
   D. all of the above

_____ 6. The sliding gauge on a shingling hatchet is used _____ .
   A. to split shakes
   B. for checking the shingle exposure
   C. to extract fasteners
   D. to trim shingles and shakes

_____ 7. The use of a power nailer on wood shingles and shakes _____ .
   A. is a tremendous time saver
   B. is not permitted
   C. could result in more time lost than gained
   D. lengthens the life expectancy of the roof

_____ 8. _____ nails are corrosion-resistant.
   A. Stainless steel
   B. Hot-dipped galvanized
   C. Aluminum
   D. all of the above

_____ 9. If staples are used to fasten wood shingles or shakes, they should be _____ .
   A. at least 20 gauge with a ⅜″ minimum crown
   B. long enough to penetrate the sheathing by at least ¼″
   C. driven with the crown across the grain
   D. long enough to penetrate the sheathing by at least ⅜″

_____ 10. If a gutter is used, overhang the wood shingle starter course _____ .
   A. ½″ past the fascia
   B. plumb with the center of the gutter
   C. 1″ past the inside edge of the gutter
   D. 1½″ past the fascia

Completion

Complete each sentence by inserting the best answer on the line near the number.

_____ 1. Place each fastener about _____ inch in from the edge of the wood shingle and not more than 1″ above the exposure line.

_____ 2. Do not allow the head of the fastener to _____ the surface of the shingle.

_____ 3. In regions of heavy snowfall, it is recommended that the starter course be _____ .

_____ 4. Joints in adjacent courses of wood shingles should be staggered at least _____ inches.

_____ 5. No joint in any _____ adjacent courses should be in alignment.

_____ 6. On intersecting roofs, do not break joints in the _____ .

_____ 7. Hip and ridge caps are usually _____ to_____ inches wide.

_____ 8. When laying wood shakes, a/an _____ consisting of strips of 18″ wide, 30# roofing felt is used.

_____ 9. Straight-split shakes should be laid with their _____ end toward the ridge.

_____ 10. It is important to regularly check the _____ of wood shakes with the hatchet handle since there is a tendency for the course to angle toward the ground.

_____ 11. Divide the total square feet of the roof area by _____ to determine the total number of squares needed for the job.

_____ 12. At standard exposure, estimate _____ pounds of nails per square of shingles.

Math Problem-Solving

Solve the following math problems.

_____ 1. What is the area for estimating gable roof shingles if the rafter is 14'-9" and the ridge is 40'-9"?

_____ 2. How many squares of wood shingles are needed to cover a roof area of 1750 square feet with 10% waste added?

_____ 3. How many extra squares of wood shingles would be needed if 50' of valley is included in the roof?

_____ 4. How many extra bundles of wood shingles would be needed for 80 lineal feet of hip and ridge?

_____ 5. If normal asphalt shingle exposure is 5" and the distance left to be covered to the center of the ridge is 41.5", what is the shingle exposure needed to make full tabs work to the ridge?

54 Flashing

Multiple Choice

Write the letter for the best answer on the line next to the number of the sentence.

_____ 1. The woven valley method of flashing _____ .
   A. is an open valley method
   B. is a closed valley method
   C. should have the course end joints occurring on the valley center line
   D. requires that sheet metal flashing be used under it

_____ 2. The closed cut method of valley flashing _____ .
   A. requires the use of fasteners at the valley's center
   B. does not require the use of 50 pound per square rolled roofing
   C. is an open valley method
   D. requires the use of asphalt cement

_____ 3. When using step flashing in a valley, each piece of flashing should be at least _____ wide, if the roof has less than a 6″ rise.
   A. 6″
   B. 12″
   C. 18″
   D. 24″

_____ 4. The height of each piece of step valley flashing should be at least 3″ more than the shingle _____ .
   A. head lap
   B. top lap
   C. exposure
   D. salvage

_____ 5. When the valley is completely flashed with the step flashing method, _____ .
   A. a 6″ wide strip of metal flashing the length of the valley is visible
   B. rolled roofing is applied over the top of it
   C. no metal flashing surface is exposed
   D. a coating of asphalt cement is applied on top of the shingles

_____ 6. The usual method of making the joint between a vertical wall and a roof watertight is the use of _____ .
   A. a saddle
   B. an apron
   C. a heavy coating of asphalt cement
   D. step flashing

_____ 7. On steep roofs between the upper side of the chimney and the roof deck, a saddle or _____ is built.
  A. heel
  B. frog
  C. cricket
  D. gusset

_____ 8. Chimney flashings are usually installed by _____ .
  A. carpenters
  B. brick masons
  C. roofers
  D. laborers

_____ 9. The upper ends of chimney flashing is _____ .
  A. fastened to the chimney with concrete nails driven through the flashing and into the mortar joints
  B. embedded in asphalt cement against the chimney
  C. bent around and mortared in between the courses of brick
  D. bent over and bedded to the shingles with asphalt cement

_____ 10. When installing flashing over a stack vent, _____ .
  A. shingle over top of the lower end of the stack vent flashing
  B. the flashing is attached with one fastener in each upper corner
  C. be sure the shingle fasteners penetrate the flashing
  D. all of the above

Completion

Complete each sentence by inserting the best answer on the line near the number.

_____ 1. _____ and zinc are expensive but high quality flashing material.

_____ 2. When reroofing, it is a good practice to replace all the _____ .

_____ 3. It is necessary to install an eaves flashing whenever there is a possibility of _____ forming.

_____ 4. Roof _____ are especially vulnerable to leaking due to the great volume of water that flows down through them.

_____ 5. When using rolled roofing as valley flashing, the first layer should be laid with its mineral surface side _____ .

_____ 6. When the end courses are properly trimmed on a valley that is 16′ long, its width at the ridge will be 6″ and at the eaves _____ inches.

_____ 7. The metal flashing in an open valley between a low-pitched roof and a much steeper one, should have a _____-inch high crimped standing seam in its center.

8. _____ valleys are those where the shingles meet in the center of the valley covering the valley flashing completely.

9. The rolled roofing required in closed valleys must be _____ pound per square or more.

10. On woven valleys, be sure that no _____ is located within 6″ of the centerline.

Math Problem-Solving

Solve the following math problems.

_____ 1. What is $\frac{7}{12}$ of 36?

_____ 2. How many inches in 24′-11″?

_____ 3. How many rows of asphalt shingles with a 5″ exposure will be needed to cover a roof with a rafter that is 16′-8″ long?

_____ 4. How many 8″ × 8″ step flashing pieces can be cut from a 2′ × 50′ roll of flashing material?

_____ 5. How many square inches in one square yard?

Discussion

Write your answer(s) on the lines below.

1. Why is it helpful to remember, when applying any roofing material, that water always runs downhill?

_____

_____

_____

55 Window Terms and Types

Multiple Choice

Write the letter for the best answer on the line next to the number of the sentence.

_____ 1. Wooden windows are primed _____ with their first coat of paint.
   A. after installation
   B. at the factory
   C. by the retailer
   D. when the siding is applied

_____ 2. Vinyl-clad wood windows _____ .
   A. come in a large variety of colors
   B. are designed to eliminate painting
   C. must be primed before installation
   D. are only available in fixed windows

_____ 3. Screens are attached to the outside of the window frame on _____ windows.
   A. awning
   B. sliding
   C. double-hung
   D. jalousie

_____ 4. _____ fixed windows are widely used in combination with other window types.
   A. elliptical
   B. half rounds
   C. quarter rounds
   D. all of the above

_____ 5. The single-hung window is similar to the double-hung window except _____ .
   A. the lower sash swings outward
   B. the parting bead is eliminated
   C. the upper sash is fixed
   D. it only has one sash

_____ 6. When the sashes are closed on double-hung windows, specially shaped _____ come together to form a weather-tight joint.
   A. parting beads
   B. meeting rails
   C. blind stops
   D. sash locks

_____ 7. The _____ window consists of a sash hinged at the side and swings outward by means of a crank or lever.
   A. casement
   B. awning
   C. hopper
   D. jalousie

_____ 8. An advantage of the casement type window is that _____ .
   A. the entire sash can be opened for maximum ventilation
   B. its system of springs and balances makes it easy to open
   C. it comes with removable sashes for easy cleaning
   D. the screens are installed on the outside of the frame

_____ 9. An awning window consists of a frame in which a sash _____ .
   A. hinged at the bottom swings inward
   B. slides horizontally left or right in a set of tracks
   C. hinged at the side swings inward
   D. hinged at the top swings outward

_____ 10. Windows installed on sloping surfaces are called _____ .
   A. skylights
   B. roof windows
   C. dormers
   D. A and B

Completion

Complete each sentence by inserting the best answer on the line near the number.

_____ 1. Wood windows, doors, and cabinets fabricated in woodworking plants are referred to as _____ .

_____ 2. The _____ is the frame that holds the glass in the window.

_____ 3. _____ are the vertical edge members of the sash.

_____ 4. Small strips of wood that divide the glass into smaller lights are called _____ .

_____ 5. Many windows come with false muntin called _____ .

_____ 6. The installation of glass in a window sash is called _____ .

_____ 7. Skylights and roof windows are generally required to be glazed with _____ .

_____ 8. To raise the R-value of insulating glass, the space between the glass is filled with _____ gas.

_____ 9. An invisible, thin _____ coating is bonded to the air space side of the inner glass of solar control insulating glass.

_____ 10. The bottom horizontal member of the window frame is called a _____ .

_____ 11. The _____ are the vertical sides of the window frame.

_____ 12. A _____ is formed where the two side jambs on side by side windows are joined together.

Identification: Windows

Identify each term, and write the letter of the best answer on the line next to each number.

_____ 1. top rail

_____ 2. bottom rail

_____ 3. stile

_____ 4. muntin

_____ 5. light

_____ 6. double-hung window

_____ 7. casement window

_____ 8. awning window

_____ 9. sliding windows

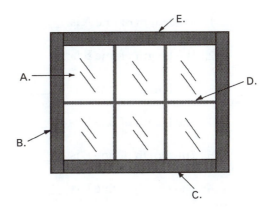

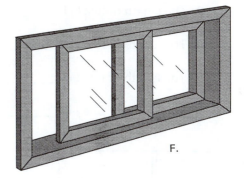

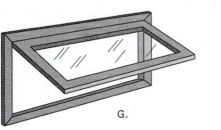

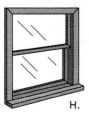

Math Problem-Solving

Provide the best answer for each of the following questions.

_____   1. The length of line A is _____ inches.

_____   2. The length of line B is _____ inches.

_____   3. The length of line C is _____ inches.

_____   4. The length of line D is _____ inches.

_____   5. The length of line E is _____ inches.

_____   6. The length of line F is _____ inches.

_____   7. The length of line G is _____ inches.

_____   8. The length of line H is _____ inches.

_____   9. The length of line I is _____ inches.

_____   10. The length of line J is _____ inches

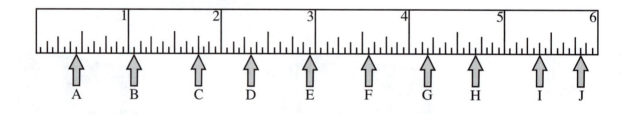

56 Window Installation and Glazing

Multiple Choice

Write the letter for the best answer on the line next to the number of the sentence.

_____ 1. Those responsible for planning the location or selection of windows need to be aware of building code requirements for minimum _____ .
  A. areas of natural light
  B. ventilation by windows
  C. window size in regard to emergency egress
  D. all of the above

_____ 2. _____ windows should not be located above porches or decks unless they are high enough to allow people to travel under them.
  A. Double- and single-hung
  B. Sliding and hopper
  C. Awning and casement
  D. Fixed

_____ 3. The builder should refer to the window _____ to determine the window style, size, manufacturer's name, and unit number.
  A. schedule
  B. agenda
  C. program
  D. plan

_____ 4. In order for the builder to better understand the construction of a particular window unit, he should refer to the _____ .
  A. local building codes
  B. Sweets register
  C. manufacturer's catalog
  D. window syllabus

_____ 5. The major purpose of applying building paper or a housewrap prior to siding application is to make the building more _____ proof.
  A. air
  B. weather
  C. moisture
  D. vapor

_____ 6. _____ will survive the longest period of exposure to the weather.
  A. Vapor retarder
  B. House wrap
  C. Polyethylene film
  D. Asphalt felt

_____ 7. To improve the performance of house wrap it is best to _____ .
   A. tape the seams with sheathing tape
   B. install the wrap after the windows are installed
   C. used a vapor retarder material
   D. all of the above

_____ 8. When using housewrap, overlap all joints by at least _____ inch(es).
   A. 1
   B. 3
   C. 6
   D. 9

_____ 9. Most windows are installed so that _____ .
   A. their tops are all at different levels
   B. their header casings lie at the same elevation
   C. the same size window is on each story
   D. the double-hung windows are on the first floor and the single-hung windows are on the second floor

_____ 10. When installing windows, _____ .
   A. remove any diagonal braces applied at the factory
   B. unlock and open the sash
   C. leave all protection blocks on the the window unit
   D. be sure to shim, level, and plumb the unit

Completion

Complete each sentence by inserting the best answer on the line near the number.

_____ 1. When it is necessary to shim the bottom of a window to level it, the shim is placed between the rough sill and the bottom end of the window's side _____ .

_____ 2. To avoid splitting the casing, all nails should be at least _____ inches back from casing's end.

_____ 3. Nails driven through the window casing should be _____ casing or common nails.

_____ 4. On vinyl-clad windows, large head _____ nails are driven through the nailing flange instead of the casing.

_____ 5. Windows are installed in masonry openings against wood _____ .

_____ 6. Tradespeople who perform the work of cutting and installing lights of glass in sashes or doors are known as _____ .

_____ 7. _____ are small triangular or diamond-shaped pieces of thin metal used to hold glass in place.

8. A light of glass is installed with its crown side _____ in a thin bed of compound against the rabbet of the opening.

9. Going over the scored line a second time with the glass cutter will _____ the glass cutter.

10. Prior to scoring a line on glass, it is recommended to brush some _____ along the line of cut.

Math Problem-Solving

Solve the following math problems using the information in the schedule.

Window Schedule				
Symbol	Quantity	Type	Manufacture Label	Rough Opening
A	1	Bow	C44 BOW	8'-1⅜" × 4'-2"
B	5	Double Hung	WPW3452	3'-6⅛" × 5'-4⅞"
C	4	Double Hung	WDH2836	2'-10⅛" × 3'-8⅞"
D	2	Casement	CN24	3'-5¼" × 4'-0½"
E	1	Sliding	244GW5046	5'-0" × 4'-6"

_____ 1. How many double hung windows are included in the schedule?

_____ 2. What is the total cost of windows D if they each cost $297.95?

_____ 3. What is the rough opening area, in square feet, of one casement window? Round answer to two decimal places.

_____ 4. What is the length of header needed for the window E if one jack is used on either side of rough opening?

_____ 5. What are the lineal feet of header needed for all double hung windows? Round up each rough opening width to the next whole inch before determining header length. Round up to the nearest whole foot.

57 Door Frame Construction and Installation

Multiple Choice

Write the letter for the best answer on the line next to the number of the sentence.

_____ 1. The top member of a door frame is called the _____ .
   A. sill
   B. head jamb
   C. bottom jamb
   D. casing

_____ 2. The part of the door frame that seals the bottom of the door is called the _____ .
   A. threshold
   B. jamb
   C. weather strip
   D. finished floor

_____ 3. Residential doors are different from commercial doors in that the residential door _____ .
   A. is smaller
   B. is larger
   C. swings towards the inside
   D. swings toward the outside

_____ 4. To plumb the side jambs of a door use a _____ .
   A. 6' level
   B. plumb bob
   C. carpenter's level and straight edge
   D. all of the above

_____ 5. A wind in a door frame is a _____ .
   A. twist
   B. rack
   C. lean
   D. rotation

_____ 6. Metal door frames set into masonry walls are _____ .
   A. set before the wall is constructed
   B. braced for movement in two directions
   C. leveled by shimming the side jambs
   D. all of the above

Completion

Complete each sentence by inserting the best answer on the line near the number.

_____ 1. Exterior doors, like windows, are manufactured in _____ plants in a wide variety of styles and sizes.

_____ 2. Many entrance doors come _____ in frames, with complete exterior casings applied, ready for installation.

_____ 3. The bottom member of an exterior door is the _____ .

_____ 4. Vertical side members of an exterior door are known as _____ .

_____ 5. In residential construction, exterior doors usually swing _____ .

_____ 6. In order to ensure an exact fit between the sill and the door, sometimes the _____ at the bottom of the door is adjustable.

_____ 7. Jamb _____ is equal to the overall wall thickness from the outside of the wall sheathing to the inside surface of the interior wall covering.

_____ 8. When setting a door frame, cut off the _____ which project(s) beyond the sill and header jamb.

_____ 9. If it is necessary to level the sill when setting the door frame, the shims are placed under the _____ .

_____ 10. A _____ is a twist in the door frame caused when the side jambs do not line up vertically with each other.

Identification: Exterior Door Components

Identify each term, and write the letter of the correct answer on the line next to each number.

_____ 1. threshold

_____ 2. head casing

_____ 3. side casing

_____ 4. head jamb

_____ 5. sill

_____ 6. band mold

_____ 7. side jamb

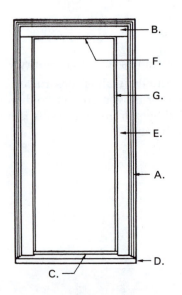

Math Problem-Solving

Solve the following math problems.

_____ 1. How many inches tall is a 6'-8" door?

_____ 2. How many feet and inches in a door that is 83" tall?

_____ 3. What is the decimal equivalent in feet of 13'-5" rounded to four decimal places?

_____ 4. What is the decimal equivalent in inches of 5'-3$\frac{11}{16}$" rounded to four decimal places?

_____ 5. How many feet-inches to the nearest sixteenth is 183.45"?

Discussion

Write your answer(s) on the lines below.

1. In buildings used by the general public that have an occupancy load exceeding fifty, building codes require exterior doors used as exits to swing outward. Why do you suppose this is so?

   _____

   _____

   _____

58 Door Fitting and Hanging

Multiple Choice

Write the letter for the best answer on the line next to the number of the sentence.

_____ 1. _____ doors are highly crafted designer doors with a variety of cut-glass designs.
  A. French
  B. Dutch
  C. High-style
  D. Sash

_____ 2. _____ doors consist of top and bottom units, hinged independently of each other.
  A. Dutch
  B. Flush
  C. French
  D. Panel

_____ 3. A panel door consists of a frame that surrounds panels of _____ .
  A. solid wood
  B. glass
  C. louvers
  D. all of the above

_____ 4. The outside vertical members of a panel door are called _____ .
  A. mullions
  B. rails
  C. stiles
  D. jambs

_____ 5. The widest of all rails in a panel door is the _____ rail.
  A. bottom
  B. top
  C. lock
  D. intermediate

_____ 6. Practically all exterior entrance doors are manufactured with a thickness of _____ inches.
  A. ¾
  B. 1
  C. 1½
  D. 1¾

_____ 7. Prehung exterior doors come _____ .
  A. already fixed and hinged in the door frame
  B. with the threshold installed
  C. with the outside casing installed
  D. all of the above

8. When installing a prehung door unit, _____ .
   A. remove the factory installed spacers between the door and the jamb before leveling and shimming
   B. avoid nailing the side jambs through the shims
   C. make sure the outside edge of the jamb is flush with the finished wall surface
   D. it is good practice to hammer the nails completely flush with the finish

9. The first step in fitting a door is to _____ .
   A. determine the side that will close against the stops on the door frame
   B. install the lockset
   C. apply the hinges
   D. attach ¼″ spacers to the door's stiles

10. Exterior doors containing lights of glass must be hung with the _____ .
    A. removable glass bead facing the exterior
    B. glass bead first removed
    C. removable glass bead facing the interior
    D. lights of glass removed

Completion

Complete each sentence by inserting the best answer on the line near the number.

1. The _____ of the door is designated as being either left-hand or right-hand.

2. A left-handed door has its hinges on the left side if you are standing on the side that swings _____ from you.

3. The process of fitting a door into a frame is called _____ .

4. The lock edge of a door must be planned on a _____ .

5. To _____ sharp corners means to round them over slightly.

6. Use _____ 4″ × 4″ hinges on 1¾″ doors 7′-0″ or less in height.

7. The recess in a door for the hinge is called the_____ , or sometimes a hinge mortise.

8. When laying out hinges, use a _____ instead of a pencil to mark the line.

9. When many doors need to be hung, a butt hinge template and a portable electric _____ are used to cut hinge gains.

10. A(n) _____ is a molding that is rabbeted on both edges and designed to cover the joint between double doors.

Math Problem-Solving

Solve the following math problems.

_____ 1. What is the sum of 6³⁄₁₆″, 5¼″, and 10⅞″?

_____ 2. Subtract 11¹³⁄₁₆″ from 20⅝″.

_____ 3. Convert ³⁄₃₂″ to a five place decimal.

_____ 4. Convert 0.368″ to the nearest sixteenth of an inch.

_____ 5. What is one-fifth of 14′-8″ in terms of feet-inches to the nearest sixteenth?

59 Installing Exterior Door Locksets

Multiple Choice

Write the letter for the best answer on the line next to the number of the sentence.

_____ 1. Decorative plates installed between the door surface and the door knob are called _____ .

    A. astragals
    B. escutcheons
    C. locksets
    D. strike plates

_____ 2. Cylindrical locksets are manufactured with _____ .

    A. round knob handles
    B. lever knob handles
    C. auxiliary deadbolts
    D. all of the above

_____ 3. To lay out the location of the holes in a door for a lockset _____ .

    A. use the paper template included in lockset packaging
    B. measure down from the top of the door
    C. use a compass to mark circles
    D. all of the above

_____ 4. To secure the door from swinging while installing locksets it is best to _____ .

    A. screw stop blocks to the floor
    B. have someone hold the door
    C. take the door off the hinges
    D. wedge the door between the bottom edge and the floor

_____ 5. The distance the center of the lockset hole is from the door edge is called _____ .

    A. offset
    B. boring distance
    C. backset
    D. centerline

Completion

Complete each sentence by inserting the best answer on the line near the number.

_____ 1. Key-in-knob is another name for _____ locksets.

_____ 2. A _____ provides additional security over a lockset alone.

_____ 3. A/An _____ lock combines deadbolt locking action with a standard key-in-knob set.

_____ 4. _____ are decorative plates of various shapes that are installed between the lock handle or knob.

_____ 5. Measuring up from the floor, the recommended distance to the centerline of a lock is usually _____ inches.

_____ 6. The _____ of the lock is the distance from the edge of the door to the center of the hole through the side of the door.

_____ 7. When boring the holes for a lockset, the hole through the _____ of the door should be bored first.

_____ 8. The typical diameter of a cylindrical lockset hole is _____ .

_____ 9. The striker plate is installed on the door _____ .

_____ 10. After the door is fitted, hung, and locked, remove all _____ and prime the door and all exposed parts of the door frame.

Math Problem-Solving

Solve the following math problems.

_____ 1. What is the average of 70, 87, and 98?

_____ 2. What is the sum of 7'-3⅛" and 2'-8⅞"

_____ 3. What is the area of a triangle with a base of 9'-3" and a height of 2'-6"? Answer in terms of square feet rounded to the nearest hundredth.

_____ 4. What is the perimeter of figure A?

_____ 5. What is the area of figure A?

Discussion

Write your answer(s) on the lines below.

1. When laying out hinges, face plates, and striker plates, it is recommended that a sharp knife be used instead of a pencil. Why is this so?

_____

_____

_____

60 Siding Types and Sizes

Multiple Choice

Write the letter for the best answer on the line next to the number of the sentence.

_____ 1. Most redwood siding is produced by mills that belong to _____ .
   A. WWPA
   B. CRA
   C. APA
   D. AHWA

_____ 2. In sidings classified as _____ , the annual growth rings must form an angle of 45° or more with the surface.
   A. flat grain
   B. vertical grain
   C. mixed grain
   D. cross grain

_____ 3. Vertical grain siding is the highest quality because it _____ .
   A. warps less
   B. takes and holds finishes better
   C. has less defects and is easier to work
   D. all of the above

_____ 4. _____ surfaces generally hold finishes longer than other types of surfaces.
   A. Smooth
   B. Flat-grained
   C. Saw-textured
   D. Double-planed

_____ 5. Knotty grade siding is divided into #1, #2, and #3 common depending on _____ .
   A. the type and number of knots
   B. its approved method of application
   C. the species of tree it comes from
   D. the thickness it is cut to

_____ 6. The best grades of redwood siding are grouped in a category called _____ .
   A. structural
   B. edifical
   C. clear
   D. architectural

_____ 7. Bevel siding is more commonly known as _____ .
   A. drop
   B. tongue and groove
   C. clapboard
   D. channel rustic

8. Most panel and lap siding is manufactured from _____ .
   A. plywood, hardboard, and cement
   B. cedar and redwood
   C. hemlock and poplar
   D. all of the above

9. Most panel siding is shaped with _____ edges for weathertight joints.
   A. back beveled
   B. chamfered
   C. shiplapped
   D. mitered

10. Lap siding comes in thicknesses from $\frac{7}{16}$ to $\frac{9}{16}$ of an inch, widths of 6, 8, and 12 inches and lengths of _____ feet.
    A. 4
    B. 8
    C. 12
    D. 16

Math Problem-Solving

Using the figure shown, solve the following math problems. Assume the sides not visible in the diagram have the same window and door area as the visible sides.

1. What is the area of all the windows? Assume front windows are all the same size.

2. What is the area of the front and back doors?

3. What is the area of two triangular gable ends?

4. What is the gross wall surface area of the house, including gables and before deducting the opening areas?

5. What is the area of the house that will actually be covered with siding?

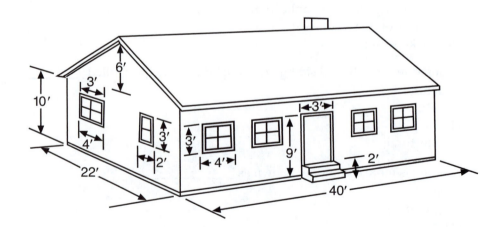

228

Name_____   Date _____

61 Applying Vertical and Horizontal Wood Siding

Multiple Choice

Write the letter for the best answer on the line next to the number of the sentence.

_____ 1. Corner boards are not usually used when wood siding is applied _____ .
   A. vertically
   B. horizontally
   C. diagonally
   D. all of the above

_____ 2. When applying tongue and groove siding vertically, fasten by _____ .
   A. face nailing into the groove edge of each piece
   B. toenailing into the groove edge of each piece
   C. toenailing into the tongue edge of each piece
   D. face nailing through the groove edge and into the tongue edge of the next piece

_____ 3. If necessary to make horizontal joints between lengths of vertical siding, _____ .
   A. a butt joint alone is acceptable
   B. install building paper beneath the joint
   C. use mitered or rabbeted end joints
   D. leave a ¹⁄₁₆″ gap to receive caulking

_____ 4. When applying short lengths of vertical siding under a window, _____ .
   A. leave an ample margin for expansion of the window
   B. apply a bead of exterior glue between the top of the siding and the window's bottom
   C. rabbet the top of the siding to fit into the weather groove on the window's bottom
   D. butt the pieces against the window's bottom

_____ 5. The last piece of vertical siding should be _____ .
   A. much narrower than the rest
   B. wider than the rest
   C. fastened with screws instead of nails
   D. as close as possible in width to the pieces previously installed

_____ 6. When installing vertical panel siding that is thicker than ½″, use _____ siding nails.
   A. 4d
   B. 6d
   C. 8d
   D. 10d

229

_____ 7. When installing horizontal or vertical panel siding, it is important that all horizontal joints be _____ .
    A. offset and lapped
    B. rabbeted
    C. flashed if a butt joint
    D. all of the above

_____ 8. When cutting the last piece of vertical siding around a window, it is best to _____.
    A. install temporary siding blocks above and below window
    B. scribe last piece to window side casing
    C. level lines from top and bottom casings
    D. all of the above

Completion

Complete each sentence by inserting the best answer on the line near the number.

_____ 1. A _____ is finish work that may be installed around the perimeter of the building slightly below the top of the foundation.

_____ 2. If the siding terminates against the soffit and no frieze is used, the joint between them is covered by a _____ molding.

_____ 3. If corner boards are not used, then horizontal siding may be _____ around exterior corners.

_____ 4. On interior corners, the siding courses may butt against a square corner _____ .

_____ 5. One of the two pieces making up an outside corner board should be narrower than the other by the thickness of the _____ .

_____ 6. Vertical siding boards are installed around windows by first _____ the piece to fit the window frame.

_____ 7. A major advantage of bevel siding over other types is the ability to vary the siding's _____ .

_____ 8. The furring strip, which is applied before the first course of siding, must be the same thickness and width as the siding _____ .

_____ 9. A _____ is often used for accurate layout of siding where it butts against corner boards, casings, and similar trim.

_____ 10. For weather-tightness when fitting siding under a window, it is important that siding fits snugly in the _____ on the underside of the window sill.

_____ 11. Bevel and Dolly Varden siding are only to be applied in the _____ position.

_____ 12. When installing vertical tongue and groove siding directly to the frame, _____ must be provided between the studs.

Math Problem-Solving

Solve the following math problems.

_____ 1. What is the actual area of the 2' wide footing shown as the shaded portion in the accompanying figure?

_____ 2. How many 1 × 10 rough cut boards will be needed to vertically side a wall section that is 33' long? Allow for a 1″ space between boards.

_____ 3. In the figure, what would be the siding exposure below the window to make the laps work out to whole boards?

_____ 4. What would be the siding exposure near the window to make the laps work out to whole boards?

_____ 5. What would be the siding exposure above the window to make the laps work out to whole boards?

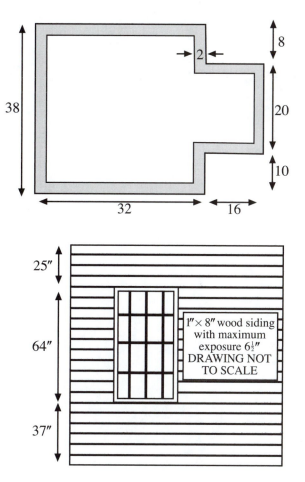

231

62 Wood Shingle and Shake Siding

Multiple Choice

Write the letter for the best answer on the line next to the number of the sentence.

_____ 1. Shingles and shakes that are rebutted and rejointed have _____ .
   A. trimmed edges
   B. parallel edge shingles (PES)
   C. S2S surfaces
   D. machine trimmed parallel edges

_____ 2. The maximum weather exposure of 24″ wood shingles installed as single course siding is _____ .
   A. 8″
   B. 10½″
   C. 16″
   D. 24″

_____ 3. The starter course of wood shingles is usually done with _____ .
   A. starter shingles
   B. undercourse shingles
   C. lower course shingles
   D. ending shingles

_____ 4. Staggered coursing of wood shingles is when the _____ .
   A. shingles are installed with varying widths
   B. butt edge rests on a straight edge
   C. butt edges in a course do not line up with each other
   D. when butt seams of many courses do not line up

_____ 5. When shingles are installed at corners they are _____ .
   A. butted to a corner board
   B. laced to each other
   C. mitered
   D. all of the above

_____ 6. How many square feet will four bundles of 16″ wood shingles cover at an exposure of 7½″?
   A. 150 square feet
   B. 140 square feet
   C. 136 square feet
   D. 100 square feet

Completion

Complete each sentence by inserting the best answer on the line near the number.

_____ 1. Rebutted and rejointed machine-grooved sidewall shakes have _____ faces.

_____ 2. Fancy butt shingles were widely used in the 19th century on _____ style buildings.

_____ 3. _____ shingles allows them to be installed with greater exposure .

_____ 4. When applying shingles to _____ , greater exposures are permitted than on roofs.

_____ 5. When more than one layer of shingles is needed, less expensive _____ shingles are used for the underlayers.

_____ 6. Untreated shingles should be spaced ⅛″ to ¼″ apart to allow for _____ and to prevent buckling.

_____ 7. Use a _____ when it is necessary to trim and fit the edges of wooden shingles.

_____ 8. If rebutted and rejointed shingles are used, no _____ should be necessary.

_____ 9. Shingles should be fastened with two nails or staples about _____ in from the edge.

_____ 10. Fasteners used on shingles should be hot-dipped galvanized, stainless steel, or aluminum and driven about 1″ above the butt line of the next _____ .

_____ 11. The side lap of shingles should be more than _____ inches .

_____ 12. On outside corners, shingles may be applied by alternately _____ each course in the same manner as applying a wood shingle ridge.

_____ 13. When double coursing wood shingles, the first course is _____ .

_____ 14. The number of squares of shingles needed to cover a certain area depends on how much of them are _____ to the weather.

_____ 15. One square of shingles will cover 100 square feet when 16″ shingles are exposed _____ inches.

Identification: Fancy Butt Shingles

Identify each term, and write the letter of the best answer on the line next to each number.

_____ 1. arrow

_____ 2. round

_____ 3. diagonal

_____ 4. octagonal

_____ 5. square

_____ 6. diamond

_____ 7. hexagonal

_____ 8. fish scale

_____ 9. half cove

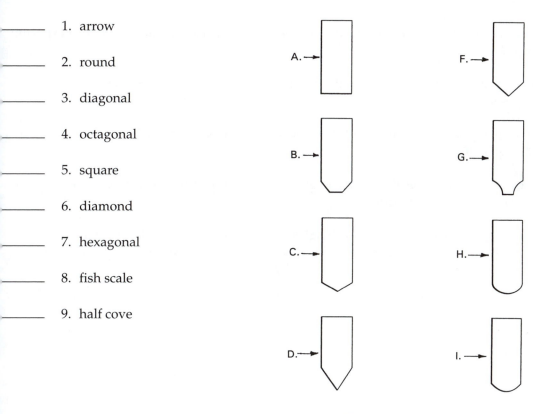

Math Problem-Solving

Solve the following math problems using information found in Figure 62-13 of textbook.

_____ 1. How many square feet of wall will one square of 16″ wood shingles cover at an exposure of 4″?

_____ 2. How many square feet of wall will one square of 24″ wood shingles cover with at an exposure of 8½″?

_____ 3. How many squares of 16″ wood shingles with an exposure of 5″ will be needed to cover a 200 square feet wall section?

_____ 4. How many squares of 18″ wood shingles with an exposure of 6½″ will be needed to cover a 295 square feet wall section?

_____ 5. How many squares of 24″ wood shingles with an exposure of 15″ installed in double coursing style will be needed to cover a 200 square feet wall section?

63 Aluminum and Vinyl Siding

Multiple Choice

Write the letter for the best answer on the line next to the number of the sentence.

_____ 1. Aluminum and vinyl siding systems are _____ .
   A. both finished with baked-on enamel
   B. similar to each other except for the material
   C. expansion resistant
   D. only available for horizontal applications

_____ 2. Aluminum and vinyl siding panels for horizontal applications are _____ .
   A. made in 8″ and 12″ widths
   B. also used for soffits
   C. made in 6″ and 9″ widths
   D. made in configurations to resemble 4, 5, or 6 courses of Dolly Varden siding

_____ 3. Panels designed for vertical application come in _____ widths and are shaped to resemble boards.
   A. 6″
   B. 8″
   C. 10″
   D. 12″

_____ 4. With changes in temperature, vinyl siding may contract and expand as much as _____ in a 12′-6″ section.
   A. ⅛″
   B. ¼″
   C. ⅜″
   D. ½″

_____ 5. Fasteners should be driven _____ the siding.
   A. very tightly against
   B. not too tightly into
   C. every 24″ along
   D. to the left of center in the slots on

_____ 6. Starter strips must be applied _____ .
   A. tightly against corner posts
   B. at vertical intervals of every four feet
   C. as straight as possible
   D. only on siding that is installed vertically

_____ 7. Install _____ across the tops and along the sides of window and door casings.

    A. starter strips
    B. corner posts
    C. undersill
    D. J-channel

_____ 8. When marking the cutout of a siding panel under a window, allow for _____ clearance under and on either side of the window.

    A. $\frac{1}{16}$″
    B. $\frac{1}{8}$″
    C. $\frac{1}{4}$″
    D. $\frac{3}{8}$″

_____ 9. _____ applied on the wall up against the soffit prior to the installation of the last course of siding panel.

    A. Starter strips are
    B. Corner posts are
    C. Undersill trim is
    D. J-channel is

_____ 10. The layout for vertical siding should be planned so that _____ .

    A. the same panel width is exposed at both ends of the wall
    B. any panels that need to have their edges cut to size are placed in the rear of the building
    C. any width adjustment made in the panels come in the center of the wall
    D. no panels are to have their edges trimmed

Identification: Vinyl Siding Systems

Identify each term, and write the letter of the best answer on the line next to each number.

_____ 1. horizontal siding starter strip

_____ 2. vertical siding or soffit

_____ 3. fascia

_____ 4. undersill trim

_____ 5. outside corner post

_____ 6. J-channel

_____ 7. inside corner post

_____ 8. perforated soffit

_____ 9. undersill finish trim

_____ 10. "F" trim

_____ 11. horizontal siding

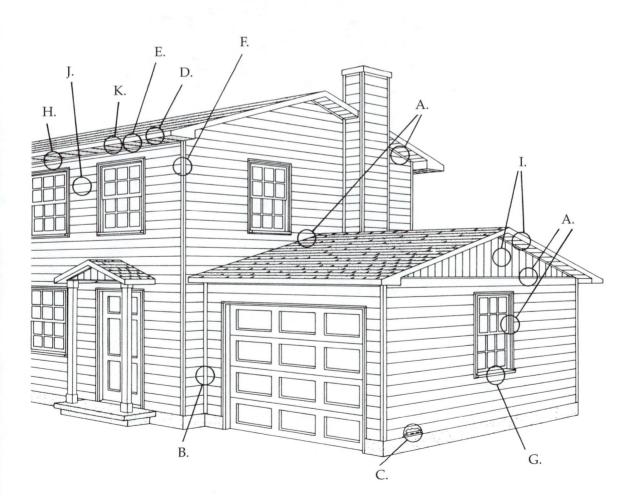

Math Problem-Solving

Solve the following math problems estimating vinyl siding parts for the figure shown. Assume the sides not visible in the diagram have the same window and door area as the visible sides.

_____ 1. How many pieces of starter strip 12' long will be needed?

_____ 2. How many pieces of J-channel 12' long will be needed to trim the perimeter of all the windows?

_____ 3. How many squares of siding will be needed if no deductions are made for the opening?

_____ 4. Using the Pythagorean theorem, determine the rake length of the siding on the gable end. Round up to the nearest whole foot?

_____ 5. How many pieces of J-channel 12' long will be needed to trim the upper edge of siding around the entire building?

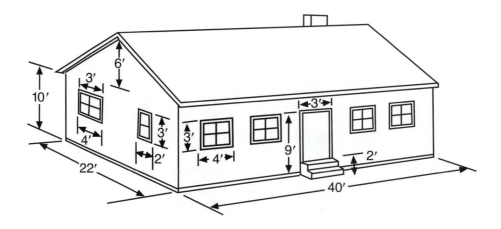

64 Cornice Terms and Design

Multiple Choice

Write the letter for the best answer on the line next to the number of the sentence.

_____ 1. The cornice that provides the least amount of weather protection to the eave is the
_____ .
   A. narrow box cornice
   B. wide box cornice
   C. snub cornice
   D. open cornice

_____ 2. The cornice type that runs up the rafter at the gable end is called _____ .
   A. rake cornice
   B. rafter cornice
   C. plancier cornice
   D. cornice return

_____ 3. Another name for the plancier is _____ .
   A. soffit
   B. frieze
   C. fascia
   D. lookout

_____ 4. The purpose of the soffit is to _____ .
   A. close off the ends of the rafters
   B. close off the underside of the rafter tails
   C. support the fascia
   D. all of the above

_____ 5. The surface where a gutter may be attached is called _____ .
   A. fascia
   B. soffit
   C. plancier
   D. frieze

_____ 6. Molding installed at the cornice as an ending point for siding is called
_____ .
   A. cornice molding
   B. frieze
   C. J-channel or undersill trim
   D. all of the above

Completion

Complete each sentence by inserting the best answer on the line near the number.

_____ 1. Lookouts are framing members used to provide a nailing surface for the _____ of a wide cornice.

_____ 2. The soffit is an ideal location for the placement of _____ .

_____ 3. The subfascia is usually of _____ inch nominal thickness.

_____ 4. The portion of the fascia that extends below the soffit is called the _____ .

_____ 5. If the frieze is not used, the _____ is used to cover the joint between the siding and the soffit.

_____ 6. The most commonly used cornice design is the _____ cornice.

_____ 7. The cornice design that lacks a soffit is the _____ cornice.

_____ 8. The least attractive cornice design is the _____ cornice.

_____ 9. A rake cornice would be found on a home with a _____ or gambrel roof.

_____ 10. A _____ is constructed to change the direction of a level box cornice to the angle of the roof.

Identification: Cornice Components

Identify each term, and write the letter of the best answer on the line next to each number.

_____ 1. rafter

_____ 2. lookouts

_____ 3. soffit

_____ 4. fascia

_____ 5. frieze

_____ 6. cornice molding

_____ 7. plate

_____ 8. roof sheathing

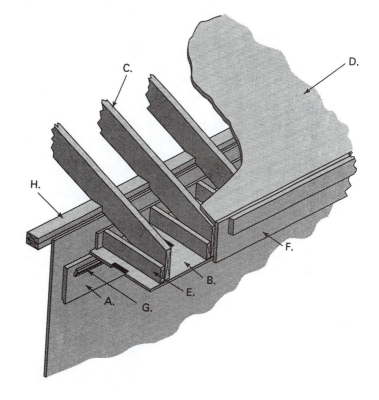

Math Problem-Solving

Provide the best answer for each of the following questions.

____ 1. The length of line A is _____ inches.

____ 2. The length of line B is _____ inches.

____ 3. The length of line C is _____ inches.

____ 4. The length of line D is _____ inches.

____ 5. The length of line E is _____ inches.

____ 6. The length of line F is _____ inches.

____ 7. The length of line G is _____ inches.

____ 8. The length of line H is _____ inches.

____ 9. The length of line I is _____ inches.

____ 10. The length of line J is _____ inches.

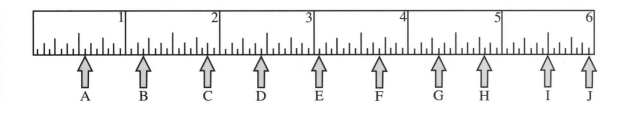

65 Gutters and Downspouts

Multiple Choice

Write the letter for the best answer on the line next to the number of the sentence.

_____ 1. The horizontal member that carries rain water from the roof is called the _____ .
- A. downspout
- B. conductor
- C. eavestrough
- D. all of the above

_____ 2. Gutters are made from _____ .
- A. wood
- B. aluminum
- C. copper
- D. all of the above

_____ 3. The fact that water runs downhill is helpful when installing _____ .
- A. gutters
- B. downspouts
- C. gutter support brackets
- D. all the above

_____ 4. The slope of the roof affects the location of the _____ .
- A. gutter
- B. gutter seams
- C. end caps
- D. all of the above

_____ 5. Water released from a gutter system should be drained away by _____ .
- A. connecting to the footing drains
- B. connecting to the foundation drains
- C. drywells or storm drains
- D. all of the above

Completion

Complete each sentence by inserting the best answer on the line near the number.

_____ 1. The _____ carries water from the gutter downward and away from the foundation.

_____ 2. No type of finish is required on _____ gutters.

_____ 3. For every _____ square feet of roof area, one square inch of gutter cross-section is needed.

_____ 4. _____ gutters can be formed to practically any length when forming machines are brought to the job site.

_____ 5. In order for water to drain toward the downspout, gutters should be installed with a pitch of about _____ inch(es) to 10 feet.

_____ 6. On long buildings, the gutter is usually _____ in the center.

_____ 7. To prevent _____ from occurring on metal gutters, be sure that the screws you are using are the same metal as the brackets.

_____ 8. Aluminum brackets may be spaced up to _____ inches on center.

_____ 9. _____ connectors are used to join the sections of metal or vinyl gutters.

_____ 10. Round corrugated galvanized iron downspouts are fastened to the wall by galvanized iron rings called _____ .

_____ 11. If downspouts are connected to the foundation, drain strainer caps must be placed over the gutter _____ .

Identification: Gutter and Downspout Parts

Identify each term, and write the letter of the best answer on the line next to each number.

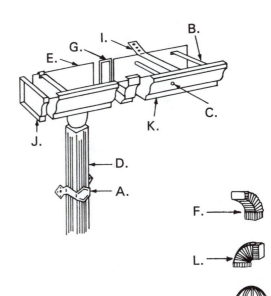

_____ 1. end cap

_____ 2. slip connector

_____ 3. strap hanger

_____ 4. gutter

_____ 5. downspout

_____ 6. fascia bracket

_____ 7. spike and ferrule

_____ 8. conductor pipe band

_____ 9. end piece

_____ 10. strainer cap

_____ 11. elbow-style A

_____ 12. elbow-style B

Math Problem-Solving

Solve the following math problems using the figure as a guide.

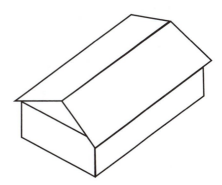

_____ 1. How many feet of gutter will be installed on the house shown if the ridgeboard is 44'-6" long?

_____ 2. How many gutter hangers would be needed if they are installed 3' OC? Add one to start.

_____ 3. How many left end caps will be needed?

_____ 4. How many downspouts would be needed?

_____ 5. How many downspout elbows would be needed?

66 Deck and Porch Construction

Multiple Choice

Write the letter for the best answer on the line next to the number of the sentence.

_____ 1. Girders are installed on the beams using _____ .
   A. post and beam metal connectors
   B. fiber tube forms
   C. post anchors
   D. joist hangers

_____ 2. When joists are hung between the girders, _____ .
   A. joist hangers are not necessary
   B. the overall length of the deck must be decreased
   C. the overall depth of the deck is decreased
   D. they must be installed with the crown down

_____ 3. When decking is run _____ to the joists, the spacing of the joists may be 24″ on center.
   A. at right angles
   B. parallel
   C. diagonal
   D. all of the above

_____ 4. If the deck is less than _____ above the ground, the supporting posts will not need to be permanently braced.
   A. 4′
   B. 6′
   C. 8′
   D. 10′

_____ 5. When applying the deck boards, _____ .
   A. lay them with the bark side down
   B. it is advisable to start at the inside edge
   C. maintain about ½″ between them
   D. make tight fitting end joints and stagger them between adjacent rows

_____ 6. A _____ board may be fastened around the perimeter of the deck with its top edge flush with the top of the deck.
   A. ledger
   B. fascia
   C. batter
   D. skirt

_____ 7. If the deck is more than 30″ above the ground, most codes require _____ high railing around the exposed sides.
   A. 24″
   B. 28″
   C. 32″
   D. 36″

_____ 8. _____ are another name for railing posts.
   A. Stanchions
   B. Balusters
   C. Preachers
   D. Rails

_____ 9. The space between the top and bottom rail may be filled with _____ .
   A. intermediate rails
   B. balusters
   C. lattice work
   D. all of the above

_____ 10. If benches are built into the deck, their seat height should be _____ above the deck.
   A. 14″
   B. 16″
   C. 18″
   D. 20″

Completion

Complete each sentence by inserting the best answer on the line near the number.

_____ 1. Lumber used in deck construction must be made either from a decay resistant species or be _____ .

_____ 2. It is the _____ of redwood and cedar that is resistant to decay.

_____ 3. If pressure-treated southern pine is used in deck construction, the grade of _____ is structurally adequate for most applications.

_____ 4. _____ is the most suitable and economical grade of California redwood for deck posts, beams, and joists.

_____ 5. All nails, fasteners, and hardware should be stainless steel, aluminum, or _____ galvanized.

_____ 6. When a deck is constructed against a building, a _____ is nailed or bolted against the wall for the entire length of the deck.

_____ 7. After the deck is applied, a _____ is installed under the siding and on top of the deck board.

_____ 8. Deck footings may require digging a hole and filling it with _____ .

_____ 9. In cold climates all deck footings must extend below the _____ .

_____ 10. All supporting posts are set on footings, then braced _____ in both directions.

Math Problem-Solving

Solve the following math problems using information from the deck in figure shown.

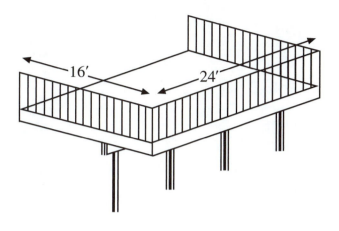

_____ 1. How many lineal feet of girder material would be needed to make a three ply built up girder?

_____ 2. How many 16″ OC joists will be needed?

_____ 3. How many lineal feet of decking boards will be needed if one board covers a width of 6″?

_____ 4. How many spindles for the railing would be needed it they are spaced 4″ apart? Neglect those spindles omitted due to the railing posts.

_____ 5. If the decking boards only covered 5¾″ in width, how many lineal feet would be needed? Round up answer to the nearest foot.

67 Fence Design and Erection

Multiple Choice

Write the letter for the best answer on the line next to the number of the sentence.

_____ 1. For the strongest fences, set the posts in _____ .
   A. clay
   B. gravel
   C. concrete
   D. mortar

_____ 2. Filling the bottom of the fence post hole with gravel _____ .
   A. reduces the required depth of the hole
   B. strengthens the post
   C. eliminates the need to brace the post
   D. helps extend the life of the post

_____ 3. When placing concrete around fence posts, _____ .
   A. form a slight depression around the post
   B. form the top so it pitches away from the post
   C. tamp it level
   D. use a very wet mix that has been allowed to partially harden

_____ 4. An alternative to embedding the fence posts in concrete is to _____ .
   A. pack gravel around the post
   B. attach the post to a metal anchor embedded in the concrete
   C. pack the post in sand
   D. all of the above

_____ 5. When installing rails on fences, keep the bottom rail at least _____ above the ground.
   A. 2"
   B. 4"
   C. 6"
   D. 8"

_____ 6. When the situation exists where the faces of the wooden posts are not in the same line as the rails, then the _____ .
   A. posts must be reset
   B. rail ends must be cut out of square to match the posts
   C. rail ends must be perfectly square
   D. rails must be slightly bowed to align them

_____ 7. If iron posts are boxed with wood, then the rails are installed _____ .
   A. in the same manner as for wood posts
   B. with special metal pipe grips
   C. by boring holes in the rails and sliding them over the posts
   D. with porcelain insulators

_____ 8. When applying spaced pickets, _____ .
       A. use a picket or a ripped piece of lumber for a spacer
       B. cut only the bottom end of the pickets when trimming their height
       C. the bottom of the pickets should not touch the ground when installed
       D. all of the above

Completion

Complete each sentence by inserting the best answer on the line near the number.

_____ 1. Because fences are not _____ structures, knotty, lower grades of lumber may be used to build them.

_____ 2. Parts of the fence that are set in the ground or exposed to constant moisture should be _____ or all-heart decay resistant wood.

_____ 3. When moisture comes in contact with inferior hardware or fasteners used on fences, corrosion results, causing unsightly _____ on the fence.

_____ 4. Placement or height of fences sometimes is restricted by _____ regulations.

_____ 5. When pickets are applied with their edges tightly together, the assembly is called a _____ fence.

_____ 6. The board-on-board fence is similar to the picket fence except the boards are _____ from side to side.

_____ 7. The _____ fence creates a solid barrier with boards or panels fitted between top and bottom rails.

_____ 8. The _____ fence permits the flow of air through it and yet provides privacy.

_____ 9. Because most post and rail designs have large _____ , they are not intended to be used as barriers to prevent passage through them.

_____ 10. Iron fence posts should be _____ or otherwise coated to prevent corrosion.

_____ 11. The first step in building a fence is to set the _____ .

_____ 12. If steep, sloping land prohibits the use of a line when setting fence posts, it may be necessary to use a _____ to lay out a straight line.

_____ 13. When building on a property line, be sure the exact locations of the _____ are known.

_____ 14. Fence posts are generally set about _____ feet apart.

Math Problem-Solving

Solve the following math problems.

_____ 1. How far apart should the centers of six evenly spaced fence posts be for a fence that is 30' long?

_____ 2. How many 3" wide pickets will be needed to build a picket fence that is 40' long?

_____ 3. How many 6" wide boards will be needed to build a board on board fence that is 32' long if no overlap occurs?

_____ 4. How far apart should the centers of eight evenly spaced fence posts be for a fence that is 32' long? Answer in terms of inches to the nearest sixteenth.

_____ 5. How many board feet of lumber will be needed if 220 pieces of 1" × 3"-6' are estimated?

Section 3: Building for Success

THE CONSTRUCTION TEAM NETWORKING PROCESS

The concept of teamwork has always been at the heart of the construction process. A network implies the use of interconnected or interrelated groups to accomplish tasks. A team network will have common goals. These shared goals become the desired outcomes that formulate and direct the team's activities. Successful teamwork outcomes are reflected in a project that is completed on or ahead of schedule, within budget, and with no or minimal worker injuries. Satisfaction for the client and all interested parties will also be achieved with the proper use of a team-centered structure.

Effective teamwork has many characteristics, but some of the most important are coordination, cooperation, scheduling, problem solving, and quality assurance. If these are implemented successfully, the goals are most generally met.

Coordination is the process of harmonizing the various people and jobs on a construction project. This action is directed through a team facilitator who might be an architect, general contractor, or field superintendent. Good coordinators have a clear picture of the entire project. They must understand the skilled trades involved, the building materials/processes, and job scheduling.

Cooperation is necessary for a team to accomplish its goals. Good interaction and communication will ensure cooperation. People find it very difficult to cooperate without trust. Flexibility is a part of cooperation and must also be possessed by team members. This helps ensure that job tasks are completed, even by alternate routes.

Good scheduling is vital if deadlines are to be met. The project coordinator has the responsibility of implementing the best delivery times for materials and the optimal time to move people on and off the work site. Planning these sequential moves will demand experience and flexibility.

Problem solving becomes the responsibility of everyone involved in the construction process. Even the best planning by experienced people will be plagued at times by the unexpected. Typical examples of the unexpected are inclement weather, delivery delays, accidents, and disagreements. As problems arise, teamwork will be implemented to solve them. Each person is called upon to share expertise and ideas in finding solutions. A problem solving process should be agreed upon by all parties in advance. Usually the general contractor has implemented a procedure based on experience.

Quality assurance is a desired outcome with the construction team. Ongoing monitoring is required by everyone to maintain product quality. The client will become more aware of job and material quality as the construction progresses. At each level of construction, the skilled workers, technicians, and laborers must be committed to achieving the highest possible standards. The key to a successful team network concept is a well prepared plan. The need for such a plan must be integrated into everyone's thinking from the beginning. People who do not buy into the team network process will cause breakdowns. These will ultimately cost the construction process time and money. Losses of this type are first felt in the profit margin. The worst possible consequence could be the lack of coordination and scheduling that might cause personal injuries to workers.

In order to ensure the welfare of the construction workers and the general public, the team network has to be accepted by everyone at the job site. Efficiency will be an outcome when full acceptance of teamwork is gained. Negative factors such as hostilities, down time, and accidents will decrease when the team network concept is accepted and implemented by everyone involved in the project. Many times the common denominator is the profit margin. All builders and construction workers will understand that teamwork will assure greater chances for a profit to be maintained on each project.

Focus Questions:
For individual or group discussion

1. The team network process in construction has its base established in working together to reach goals. What might be some good reasons to consider communicating or networking well with others in the construction industry?

2. What do you feel are necessary components in a network system approach to construction? What understandings have to be present for this system to be functional?
3. Can other components become part of a network system when considering all of the varied types of trades represented in the world of construction? What are they?
4. By concentrating on any one of the team network components, how can you defend its need?

68 Gypsum Board

Multiple Choice

Write the letter for the best answer on the line next to the number of the sentence.

_____ 1. Another name for gypsum board is _____ .
   A. drywall
   B. sheetrock
   C. plasterboard
   D. all of the above

_____ 2. Gypsum board is composed of _____ .
   A. compressed paper coated with a gypsum surface
   B. wood fibers coated with a gypsum surface
   C. 100% gypsum
   D. a gypsum core encased in paper

_____ 3. The long edges of the most commonly used gypsum board panels are _____ .
   A. tongue and groove
   B. rabbeted
   C. tapered
   D. mitered

_____ 4. Eased-edge gypsum board has a special _____ .
   A. tapered rounded edge
   B. thickened rounded edge
   C. tapered square edge
   D. thickened square edge

_____ 5. Type X gypsum board is typically known as _____ .
   A. fire code board
   B. gypsum lath
   C. red board
   D. backing board

_____ 6. Water-resistant gypsum board is easily recognized by its distinctive _____ face.
   A. brown
   B. green
   C. yellow
   D. orange

_____ 7. Blue board is the common name for _____ .
   A. gypsum lath
   B. veneer plaster base
   C. aluminum foil-backed gypsum board
   D. predecorated panels

_____ 8. The most commonly used thickness of gypsum board for walls and ceilings in manufactured housing is _____ .
   A. ¼″
   B. ⁵⁄₁₆″
   C. ⅜″
   D. ½″

_____ 9. _____ is used extensively as a base for ceramic tile.
   A. Coreboard
   B. Linerboard
   C. Brown board
   D. Cement board

_____ 10. Staples are an approved fastener for gypsum panels _____ .
   A. in all circumstances
   B. on base layers of multilayer applications when they penetrate the supports at least ⅝″
   C. only along tapered edges on regular gypsum board
   D. in warmer climates only

Completion

Complete each sentence by inserting the best answer on the line near the number.

_____ 1. The heads on gypsum board nails must be at least _____ inch in diameter.

_____ 2. A drywall hammer has a face that is _____ .

_____ 3. Care should be taken to drive nails at a right angle into gypsum board panels to prevent breaking the _____ .

_____ 4. Type _____ drywall screws are used for fastening into wood.

_____ 5. Type _____ drywall screws are used for fastening into gypsum backing boards.

_____ 6. Type _____ drywall screws are used for fastening into heavier gauge metal framing.

_____ 7. It is especially important to wear eye protection when driving drywall screws into _____ framing.

_____ 8. For bonding gypsum board directly to supports, special drywall _____ adhesive or approved construction adhesive is used.

_____ 9. Caution must be exercised when using some types of drywall adhesives that contain a _____ solvent.

_____ 10. When laminating gypsum boards to each other, no supplemental fasteners are needed if _____ adhesives are used.

Math Problem-Solving

Solve the following math problems. Round answers to the nearest whole number unless directed otherwise.

_____ 1. What is the radius of a circle with a diameter of 13'-7¾"? Answer in inches rounded to nearest eighth.

_____ 2. What is the area of a circle if the diameter is 62?

_____ 3. What is the circumference of a circle if the radius is 77?

_____ 4. How many square feet in a semicircle with a diameter of 13'-6"?

_____ 5. How many feet of fence are needed to surround one-half of a circular pool that is 18' in diameter?

69 Single-Layer and Multilayer Drywall Application

Multiple Choice

Write the letter for the best answer on the line next to the number of the sentence.

_____ 1. Drywall should be delivered to the job site _____ .
   A. as soon as the roof is on
   B. when all the rough carpentry is complete
   C. shortly before installation begins and the building is watertight
   D. anytime after construction begins

_____ 2. Drywall should be stored _____ .
   A. on its edge leaning against a wall
   B. under cover and stacked flat on supports
   C. on three supports at least 2″ wide
   D. under cover and on its edge

_____ 3. The _____ is a tool for making cuts in gypsum board.
   A. utility knife
   B. drywall saw
   C. electric drywall cutout tool
   D. all of the above

_____ 4. Stud edges that are to have gypsum panels installed on them must not be out of alignment more than _____ to adjacent studs.
   A. $\frac{1}{16}$″
   B. $\frac{1}{8}$″
   C. $\frac{1}{4}$″
   D. $\frac{3}{8}$″

_____ 5. When the single nailing method is used, nails are spaced a maximum of _____ on center for walls.
   A. 4″
   B. 6″
   C. 8″
   D. 10″

_____ 6. With double-nailing, the first nail must be _____ after driving the second nail of each set.
   A. removed
   B. reseated
   C. partially withdrawn
   D. set flush to the surface of the panel

_____ 7. When fastening drywall ceilings with screws to framing members that are 16″ on center, the screws should be spaced _____ on center.
   A. 4″
   B. 8″
   C. 12″
   D. 16″

8. When applying adhesives to studs where two panels are joined, _____ .
   A. apply one straight bead to the centerline of the stud
   B. zigzag the bead across the stud's centerline
   C. apply two parallel beads to either side of the centerline
   D. cover the entire stud with adhesive

9. On ceilings where adhesive is used, the field is fastened at about _____ intervals.
   A. 8"
   B. 12"
   C. 16"
   D. 24"

10. The reason for prebowing gypsum panels is to _____ .
    A. compensate for misaligned studs
    B. eliminate the need for adhesives
    C. eliminate the need for fasteners at the top and bottom plates
    D. reduce the number of supplemental fasteners required

11. _____ are supports made in the form of a "T" that are used to help hold ceiling drywall panels in position when fastening.
    A. Deadmen
    B. Strong backs
    C. Ledger boards
    D. Sleepers

12. If a fastener misses a support, _____ .
    A. continue driving it until it is flush with the surface of the panel
    B. remove it and dimple the hole
    C. set it beneath the panel's surface
    D. drive its head clear through the panel, then patch the hole

Completion

Complete each sentence by inserting the best answer on the line near the number.

1. When walls are less than 8'-1" high, wallboard is usually installed _____ .

2. The best way to minimize end joints when hanging drywall is to use the _____ panel of drywall as possible.

3. End joints should not fall on the same _____ as those on the opposite side of the partition.

4. To be less conspicuous, end joints should be as far from the _____ of the wall as possible.

5. When using a drywall cutout tool, care must be taken not to plunge too deeply and contact _____ .

6. When walls are more than 8'-1" high, _____ application of wallboard is more practical.

7. The _____ method of drywall application helps prevent nail popping and cracking where walls and ceilings meet.

8. When applying gypsum panels to curved surfaces, _____ the panels enable them to bend easier.

9. _____ gypsum board and cement board panels are used in bath and shower areas as bases for the application of ceramic tile.

10. With multilayer application of gypsum board, the joints of the face layer are offset at least _____ inches from the joints in the base layer.

Math Problem-Solving

Solve the following math problems.

1. What is the volume of a cube with a side measuring 10″?

2. What is the volume of the rectangular solid with a base measuring 4′ × 8′ that is 12′ high?

3. What is the volume of concrete for a 4″ thick slab that is 9′ × 27′? Answer in cubic yards.

4. How many cubic feet of gypsum is needed to make two sheets of ½″ thick drywall that measures 4′ × 16′?

5. What is the volume of the figure shown?

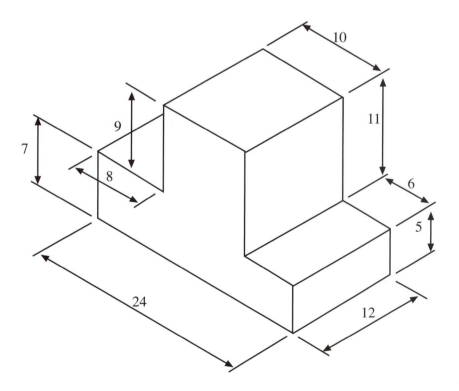

70 Concealing Fasteners and Joints

Multiple Choice

Write the letter for the best answer on the line next to the number of the sentence.

_____ 1. For 24 hours before, during, and for at least 4 days after the application of joint compound, the temperature should be maintained at a minimum of _____ .

A. 40°
B. 50°
C. 60°
D. 70°

_____ 2. Joints between panels that are ¼" or more should be _____ .

A. prefilled with compound
B. filled with insulation
C. moistened prior to filling
D. primed with a latex primer

_____ 3. When embedding the tape in compound, be sure _____ .

A. to center the tape on the joint
B. there are no air bubbles under the tape
C. there is not over ½₂" of joint compound under the edges
D. all of the above

_____ 4. Immediately after embedding the tape in compound, _____ .

A. moisten it with a fine spray of water
B. wipe it with a damp sponge
C. lift the edges and apply additional compound
D. apply another thin coat of compound to the tape

_____ 5. The term spotting in drywall work refers to _____ .

A. the application of compound to conceal fastener heads
B. air bubbles under the tape
C. stains that appear as a result of using contaminated compound
D. compound that drops to the floor during application

_____ 6. When applying compound to corner beads, _____ .

A. use the nose of the bead to serve as a guide for applying the trim
B. apply the compound about 6" wide from the nose of the bead to a feather edge on the wall
C. each subsequent finishing coat is applied about 2" wider than the previous one
D. all of the above

_____ 7. The second coat of compound is sometimes called the _____ coat.

A. fill
B. pack
C. plug
D. supply

_____ 8. Professional drywall finishers _____ .
   A. sand between each coat
   B. moisten the compound before sanding
   C. rarely sand any excess between coats
   D. only have to apply one coat

_____ 9. When finishing interior corners, _____ .
   A. a setting compound is usually applied to one side only of each corner
   B. joint reinforcing tape is not necessary
   C. the first coat is applied approximately 16″ wide
   D. all of the above

_____ 10. For every 1,000 square feet of drywall area, _____ lbs. of conventional joint compound are needed.
   A. 105
   B. 115
   C. 135
   D. 155

Completion

Complete each sentence by inserting the best answer on the line near the number.

_____ 1. Joints are reinforced with _____ .

_____ 2. Exterior corners are reinforced with _____ .

_____ 3. Taping compound is used to embed and adhere tape to the board over the _____ .

_____ 4. Second and third coats over tapped joints are covered with a _____ compound.

_____ 5. An all purpose compound may be convenient, but it lacks the _____ and workability that a two-step compound system has.

_____ 6. Setting-type joint compounds are only available in a _____ form.

_____ 7. Setting type joint compounds permit the _____ of drywall interiors in the same day.

_____ 8. To simplify its application to corners, joint tape has a _____ along its center.

_____ 9. Glass fiber mesh tape is available with a plain back or with a(n) _____ backing for quick application.

_____ 10. Instead of using fasteners on corner beads, a(n) _____ tool may be used to lock the bead to the corner.

_____ 11. Control joints are placed in large drywall areas to relieve _____ from expansion and contraction.

Math Problem-Solving

Solve the following math problems for a room that measures 18′ × 24′ by 9′ high.

_____ 1. What is the wall area for this room? Neglect openings.

_____ 2. What is total drywall area for a flat ceiling and walls in this room?

_____ 3. How many sheets of 4′ × 12′ drywall would be needed to cover the walls and ceiling?

_____ 4. How many feet of joint tape would be needed? Round up to the nearest 500 feet.

_____ 5. How many 5-gallon pails of joint compound would be needed?

Name _____   Date _____

71 Types of Wall Paneling

Multiple Choice

Write the letter for the best answer on the line next to the number of the sentence.

_____ 1. Sheet paneling is made from _____ .
   A. plywood
   B. particle board
   C. hardboard
   D. all of the above

_____ 2. Vinyl covered paneling is an example of _____ .
   A. prefinished plywood
   B. hardwood paneling
   C. softwood paneling
   D. all the above

_____ 3. Vertical plastic laminate thickness is _____ .
   A. ⅟₃₂″
   B. ⅟₁₆″
   C. ⅛″
   D. all of the above

_____ 4. Plastic laminate is installed using _____ .
   A. plastic adhesive
   B. plastic cement
   C. contact cement
   D. construction adhesive

_____ 5. A wood species used for solid wood paneling that has a dark wood tone is _____ .
   A. birch
   B. cedar
   C. cherry
   D. fir

_____ 6. The average moisture content of the air in the Rocky Mountain region is about

   _____ .
   A. 4 percent
   B. 6 percent
   C. 8 percent
   D. 11 percent

Completion

Complete each sentence by inserting the best answer on the line near the number.

_____ 1. The most widely used kind of sheet paneling is _____ .

_____ 2. Less expensive plywood paneling is prefinished with a _____ wood grain or other design on a vinyl covering.

_____ 3. The most commonly used length of paneling is the _____ foot length.

_____ 4. Matching _____ is available to cover panel edges, corners, and joints.

_____ 5. When exposed fastening is necessary, matching colored ring-shanked nails called _____ are used.

_____ 6. _____ is a hardboard panel with a baked-on plastic finish that is embossed to simulate ceramic wall tile.

_____ 7. Commonly used thicknesses of hardboard paneling range from ⅛″ to _____ inch(es).

_____ 8. Particleboard panels must only be applied to a wall backing that is _____ .

_____ 9. Unfinished particleboard paneling made from aromatic cedar chips is used to cover walls in _____ .

_____ 10. Kitchen cabinets and countertops are widely surfaced with plastic _____ .

_____ 11. _____-type laminate is used on cabinet sides and walls.

_____ 12. Regular or standard laminate is generally used on _____ surfaces.

_____ 13. Once a sheet of laminate comes in contact with the adhesive, it can no longer be _____ .

_____ 14. Most board paneling comes in a _____-inch thickness.

_____ 15. To avoid shrinkage, board paneling, like all interior finish, must be dried to a _____ content.

Identification: Solid Wood Paneling Patterns

Identify each term, and write the letter of the best answer on the line next to each number.

_____ 1. matched and eased channel

_____ 2. matched, edge and center grooved

_____ 3. matched and V-grooved

_____ 4. pickwick

_____ 5. channel rustic

_____ 6. shiplapped and V-grooved

_____ 7. tongue and grooved

_____ 8. matched, V-grooved, and beaded

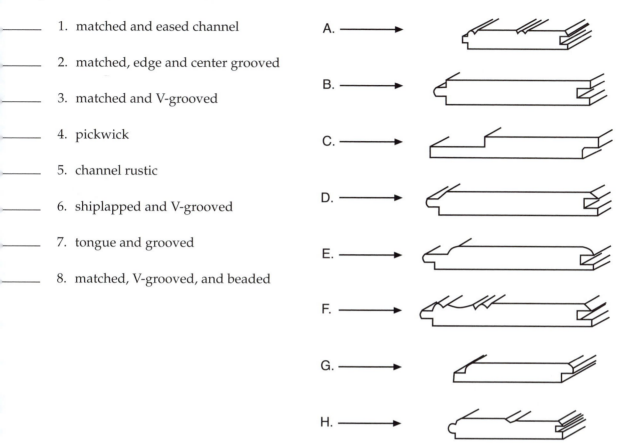

Math Problem-Solving

Solve the following math problems. Round each answer to the nearest whole number.

_____ 1. What is the volume of the figure?

_____ 2. What is the volume of 5 cylinders with a height of 10 and a radius of 4?

_____ 3. What is the volume of a cylindrical hole that measures 24′-9″ across and 5′ deep?

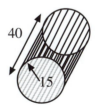

_____ 4. If there are 7.8 gallons in one cubic foot of water, how many gallons of water in a circular pool with measurements of 18′ across and 4′ deep?

_____ 5. If there are 62.4 pounds in one cubic foot of water, how many pounds would the water weigh in a circular pool that measures 18′ across and 4′ deep?

72 Application of Wall Paneling

Multiple Choice

Write the letter for the best answer on the line next to the number of the sentence.

_____ 1. Sheet paneling is usually applied to walls with the long edges _____ .
   A. vertical
   B. horizontal
   C. diagonal
   D. tight against the ceiling

_____ 2. The installation of a gypsum board base layer beneath sheet paneling is done so that _____ .
   A. the wall is stronger and more fire resistant
   B. sound transmission is deadened
   C. there will be a rigid finished surface for application of paneling
   D. all of the above

_____ 3. Before the installation of paneling to masonry walls, _____ .
   A. contact adhesive must be applied to the masonry
   B. furring strips must be applied to the masonry
   C. the wall must be parged with portland cement
   D. the back of the paneling sheet must be sealed

_____ 4. Paneling edges must fall _____ .
   A. midway between studs
   B. in line with the left side of the stud
   C. in line with the right side of the stud
   D. on stud centers

_____ 5. Panels are usually fastened with _____ .
   A. screws
   B. finish nails
   C. contact adhesive
   D. color pins and adhesive

_____ 6. Paneling that only covers the lower portion of the wall is called _____ .
   A. wainscoting
   B. louvering
   C. ledgering
   D. garthcoting

_____ 7. When cutting to a scribed line that must be made on the face side of prefinished paneling, it is recommended that a _____ be used.
   A. circular saw
   B. hand ripsaw
   C. fine-toothed hand crosscut saw
   D. saber saw

_____ 8. When using adhesive on paneling, _____ .
   A. apply beads about 4″ long and 16″ apart where panel edges and ends make contact
   B. do not allow the panel to move once it has made contact with the wall
   C. apply a ⅛″ continuous bead to the intermediate studs only
   D. after the initial contact is made between the wall and the panel, the sheet is pulled a short distance away from the wall and then pressed back into position

_____ 9. When cutting openings for wall outlets, _____ .
   A. a saber saw may be used if the cut is made from the back of the panel
   B. a circular saw may be used
   C. the openings may be cut oversized to allow for error
   D. a ripsaw is used

_____ 10. When applying paneling to exterior corners, _____ corner molding may be used.
   A. wood
   B. metal
   C. vinyl
   D. all of the above

Completion

Complete each sentence by inserting the best answer on the line near the number.

_____ 1. For vertical application of board paneling to a frame wall, _____ must be provided between the studs.

_____ 2. Prior to its application, board paneling should stand against the walls around the room to allow it to adjust to room temperature and _____ .

_____ 3. If tongue and groove board paneling is used, tack the first board in a plumb position with the _____ edge in the corner.

_____ 4. When fastening tongue and groove board paneling, blind nail into the _____ only.

_____ 5. If board paneling is of uniform width, the _____ of the first board must be planned to avoid ending with a small strip.

_____ 6. When fastening the last piece of vertical board siding, the cut edge goes in the _____ .

_____ 7. Blocking between studs on open walls is not necessary when siding is applied _____ .

_____ 8. Specially matched _____ is used between panels and on interior and exterior corners of plastic laminates that are prefabricated to plywood sheets.

Math Problem-Solving

Solve the following math problems for a room that measures 16′ × 20′ and has 8′ high walls.

_____ 1. What is the perimeter of the room?

_____ 2. How many sheets of 4′ × 8′ paneling would be needed for the walls of the room if it also has one door, one window, and a fireplace?

_____ 3. What is the wall area of room?

_____ 4. Using the table in Figure 72-16 of the text, how many board feet of solid 1 × 6 T&G wood paneling will be needed for the walls? Round up the answer to the nearest whole number and neglect any waste factor.

_____ 5. How many extra board feet should be ordered for the walls if 5% waste factor is used? Round up answer to whole board feet.

Discussion

Write your answer(s) on the lines below.

1. Describe the process of estimating the number of sheets of paneling needed to do a room.

_____

_____

_____

73 Ceramic Tile

Multiple Choice

Write the letter for the best answer on the line next to the number of the sentence.

_____ 1. The proper backing for ceramic wall tile when installed in wet locations is _____ .

    A. cement board
    B. MR gypsum board
    C. plywood
    D. all of the above

_____ 2. The most commonly used wall tiles are nominal size _____ squares, in about ¼″ thickness.

    A. 1″ and 3″
    B. 2″ and 5″
    C. 3½″ and 4½″
    D. 4″ and 6″

_____ 3. When tile is installed around a tub that lacks a showerhead, it should extend a minimum of _____ above the rim.

    A. 4″
    B. 6″
    C. 8″
    D. 10″

_____ 4. Around tubs with showerheads, the tile should extend a minimum of _____ above the rim or 6″ above the showerhead, whichever is higher.

    A. 3′
    B. 4′
    C. 5′
    D. 6′

_____ 5. Before beginning the application of ceramic wall tile, the width of the _____ must be determined.

    A. border tile
    B. field tile
    C. application trowel
    D. tile saw kerf

_____ 6. When troweling adhesive to the wall in preparation for tile application, _____ .

    A. it is best to go on the heavy side
    B. a regular straight edge trowel is to be used
    C. overlap the coverage area with the adhesive
    D. be sure the trowel is the one recommended by the manufacturer

_____ 7. Whole tiles that are applied to the center of the wall are called _____ tiles.

   A. bullnose
   B. counter
   C. field
   D. court

_____ 8. A hand-operated ceramic tile cutter works in a manner that is similar to a _____ .

   A. glass cutter
   B. masonry saw
   C. rasp
   D. coping saw

_____ 9. After all tile has been applied, the joints are filled with _____ .

   A. portland cement
   B. tile grout
   C. tile mastic
   D. contact cement

_____ 10. After the joints are partially set up but not completely hardened, they then must be
   _____ .

   A. pointed
   B. checked
   C. glazed
   D. pitched

Math Problem-Solving

Solve the following math problems.

_____ 1. What is the one-half of 13⅜″?

_____ 2. If 6″ × 6″ ceramic tile is used on a 52¾″-wide back wall of a shower, how many full tiles should be used in a row of tile?

_____ 3. What is the measurement of the width of a border in question 2?

_____ 4. How many 4″ × 4″ tiles would be ordered to cover a wall area of 90 SF? Use a 5% waste factor and round up to the nearest tile.

_____ 5. How many 4″ × 6″ tiles would be needed to cover 75 SF with a 5% waste factor? Round up to the nearest tile?

74 Suspended Ceilings

Multiple Choice

Write the letter for the best answer on the line next to the number of the sentence.

_____ 1. A main advantage of using a suspended ceiling is that the space above it can be utilized for _____ .
   A. recessed lighting
   B. duct work
   C. pipes and conduit
   D. all of the above

_____ 2. Suspended ceiling systems consist of panels that are _____ .
   A. stapled into firing strips
   B. applied with adhesive
   C. laid into a metal grid
   D. tongue and groove panels that interlock into each other

_____ 3. L-shaped pieces that are fastened to the wall to support the ends of main runners and cross tees are called _____ .
   A. wall ties
   B. wall angles
   C. support angles
   D. runner supports

_____ 4. Main runners are shaped in the form of a(n) _____ .
   A. L
   B. T
   C. upside down L
   D. upside down T

_____ 5. Slots are punched in the side of the runners at _____ intervals to receive cross tees.
   A. 4″
   B. 8″
   C. 12″
   D. 16″

_____ 6. The primary supports for the ceiling's weight are the _____ .
   A. main runners
   B. cross tees
   C. ceiling panels
   D. wall ties

_____ 7. Panels come in 2′ × 2′ and 2′ × 4′ sizes with square and _____ edges.
   A. rounded
   B. beveled
   C. concaved
   D. rabbeted

_____ 8. Main runners are usually spaced _____ apart.
    A. 2′
    B. 4′
    C. 6′
    D. 8′

_____ 9. The first parts of the ceiling grid system to be installed are the _____ .
    A. main runners
    B. cross tees
    C. wall angles
    D. wall ties

_____ 10. A suspended ceiling must be installed with at least _____ clearance below the lowest air duct, pipe, or beam for enough room to insert ceiling panels in the grid.
    A. 1″
    B. 3″
    C. 5″
    D. 8″

Completion

Complete each sentence by inserting the best answer on the line near the number.

_____ 1. Screw eyes are to be installed not over _____ feet apart.

_____ 2. Screw eyes must be long enough to penetrate wood joists by at least _____ inch(es).

_____ 3. For residential work, _____ gauge hanger wire is usually used.

_____ 4. About 6″ of hanger wire is inserted through the screw eye and then securely wrapped around itself _____ times.

_____ 5. A cross tee line must be stretched across the short dimension of the room to line up the _____ in the main runners.

_____ 6. Cross tees are installed by inserting the tabs on the ends into the slots in the _____ .

_____ 7. Ceiling tiles that are cut to fit along the perimeter of the room are called _____ tiles.

_____ 8. Panels are cut with a sharp _____ .

_____ 9. Always cut ceiling panels with their finished side placed _____ .

_____ 10. To find the number of wall angles needed in a room, divide the perimeter of the room by _____ .

Math Problem-Solving

Solve the following math problems for a room that measures 17'-9" × 23'-3" and has a 2' × 4' grid suspended ceiling installed.

_____ 1. How many pieces of wall angle should be ordered for the room?

_____ 2. If the main runners run parallel to the length of the room, how many rows of main tees should be installed?

_____ 3. How many pieces of main tees 12' long should be ordered?

_____ 4. What is the length of the border tile?

_____ 5. What is the width of the border tile?

75 Ceiling Tile

Multiple Choice

Write the letter for the best answer on the line next to the number of the sentence.

_____ 1. Ceiling tiles are made of _____ .
   A. wood fibers
   B. hardboard
   C. plywood
   D. all of the above

_____ 2. Ceiling tile is a rectangle with edges that are _____ .
   A. tongue and grooved with four fastening flanges
   B. tongue and grooved with two fastening flanges
   C. grooved on all sides
   D. tongued on all sides

_____ 3. The width of a border tile in a room that measures 10'-9" wide is _____ .
   A. 9"
   B. 9½"
   C. 10"
   D. 10½"

_____ 4. The actual number of square feet in a room that measures 10'-9" × 12'-3" is _____ .
   A. 131.7 square feet
   B. 132 square feet
   C. 132.2 square feet
   D. 143.0 square feet

_____ 5. The estimated number of 12" × 12" ceiling tiles needed for room that measures 10'-9" × 12'-3" is _____ .
   A. 120
   B. 131
   C. 132
   D. 143

_____ 6. The ceiling tiles installed first are _____ .
   A. in the middle of the room
   B. along the long wall
   C. along the short wall
   D. border tiles along two walls making a corner

Completion

Complete each sentence by inserting the best answer on the line near the number.

_____ 1. _____ fiber tiles are the lowest in cost.

_____ 2. _____ fiber tiles are used when a more fire-resistant tile is required.

_____   3. The most popular size square tile is the _____ inch.

_____   4. If the short walls of a room measure 12'-6", then the border tiles along the long walls of the room would measure _____ .

_____   5. When tiles are not being applied with adhesive to an existing ceiling, then _____ must be installed to fasten the tiles to.

_____   6. Ceiling joists are usually _____ inches on center.

_____   7. If the size of the tiles are 12", then the furring strips must be installed _____ inches on center.

_____   8. Furring strip fasteners must penetrate at least _____ inch(es) into the joist.

_____   9. Prior to installing, ceiling tiles should be allowed to adjust to normal interior room conditions for _____ hours.

_____   10. To help prevent fingerprints and smudges on the finished ceiling, some carpenters sprinkle _____ on their hands.

_____   11. On border tiles, all _____ edges should go against the wall.

_____   12. The correct number of ½" or ⁹⁄₁₆" staples to use in 12" square ceiling tiles is _____ .

Math Problem-Solving

Solve the following math problems. Round answers to the nearest two decimal places unless otherwise directed.

_____   1. Evaluate the solution of $\sqrt{(8^2 + 11^2)}$ to the nearest hundredth.

_____   2. Evaluate the solution of $\sqrt{(36^2 + 80^2)}$ to the nearest hundredth.

_____   3. What is the hypotenuse of a right triangle with sides measuring 11'-6" and 9'-9" in terms of decimal feet?

_____   4. What is the rake length, side C, in the figure shown in terms of inches to the nearest sixteenth?

_____   5. If $A^2 + B^2 = C^2$, what does A^2 equal, in terms of B^2 and C^2?

C   6'

24'-0"

76 Description of Interior Doors

Multiple Choice

Write the letter for the best answer on the line next to the number of the sentence.

_____ 1. Interior doors are typically manufactured in thicknesses of _____ .
   A. 2'-6"
   B. 3'-0"
   C. 1⅜"
   D. 1¾"

_____ 2. Flush doors are made _____ .
   A. primarily for bathrooms
   B. with solid or hollow cores
   C. with horizontal slats
   D. of a frame with smooth center pieces

_____ 3. French doors are made _____ .
   A. only in France
   B. of a frame with center panels
   C. with smooth plywood surfaces
   D. with panes of glass

_____ 4. Single-acting doors are designed to _____ .
   A. swing in one direction to open
   B. swing in two directions to open
   C. be used once
   D. not allow for the door opening to be fully open

_____ 5. The door type that does not allow complete access to the opening is the _____ .
   A. double-acting door
   B. bifold door
   C. bypass door
   D. pocket door

_____ 6. A pocket door is designed _____ .
   A. for exterior applications
   B. to have a mail slot cut into it
   C. to disappear into a wall
   D. all of the above

_____ 7. A door that operates with two doors hinged together is the _____ .
   A. folding door
   B. bypass door
   C. bifold door
   D. pocket door

_____ 8. The style of door that has two double-acting doors is called the _____ .
  A. tavern doors
  B. café doors
  C. western doors
  D. bar doors

Completion

Complete each sentence by inserting the best answer on the line near the number.

_____ 1. The interior _____ door is used when a less expensive smooth surfaced door with a plain appearance is desired.

_____ 2. Most interior residential doors are manufactured in a _____-inch thickness.

_____ 3. The most common height of manufactured doors is _____ .

_____ 4. Door widths range from 1'-0" to _____ in increments of two inches.

_____ 5. Solid core doors are generally used on _____ doors.

_____ 6. The _____ is the thin plywood that covers the frame and mesh of a hollow core door.

_____ 7. Doors that swing in both directions are known as _____ doors.

_____ 8. _____ doors are not practical in openings less than 6' wide.

_____ 9. _____ doors require more time and material to install than other kinds of doors.

_____ 10. Bifold doors on the jamb side swing on _____ installed at the top and bottom.

Matching

Write the letter for the best answer on the line near the number to which it corresponds.

_____ 1. lauan plywood   A. obstruct vision but permit the flow of air

_____ 2. french doors   B. ride on rollers in a double track

_____ 3. louver doors   C. used extensively for flush door skins

_____ 4. bypass doors   D. may contain from 1 to 15 lights of glass

_____ 5. pocket doors   E. hung in pairs that swing in both directions

_____ 6. café doors   F. only the door's lock edge is visible when opened

Math Problem-Solving

Solve the following math problems.

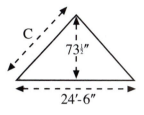

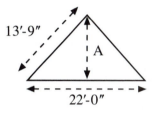

_____ 1. Evaluate the solution of $\sqrt{(18^2 + 12^2)}$ to the nearest hundredth.

_____ 2. Evaluate the solution of $\sqrt{(6^2 + 8^2)}$ to the nearest hundredth.

_____ 3. What is the hypotenuse of a right triangle with sides measuring 11'-7" and 9'-5" in terms of feet-inches to the nearest sixteenth?

_____ 4. What is the rake length, side C, in the figure shown? Answer in terms of inches to the nearest sixteenth.

_____ 5. What is the height of a gable roof if the rake length is 13'-9" and the span is 22'? Answer in terms of inches.

Discussion

Write your answer(s) on the lines below.

1. What are some of the special conditions present in heavy commercial buildings that would require heavier and larger interior doors than those used in homes?

_____

_____

_____

77 Installation of Interior Doors and Door Frames

Multiple Choice

Write the letter for the best answer on the line next to the number of the sentence.

_____ 1. The first step in making an interior door frame is to _____ .
   A. install the stop
   B. check the width and height of the rough opening
   C. rabbet the edge of the jamb stock
   D. install the header jamb

_____ 2. The rough opening width for single acting swinging doors should be the door width plus _____ .
   A. the thickness of the jamb, plus 1″
   B. double the door thickness
   C. double the side jamb thickness, plus ½″
   D. triple the side jamb thickness, plus ¾″

_____ 3. The rough opening height of the door should be the door's height plus _____ , plus the thickness of the finish floor, plus the desired clearance under the door.
   A. the thickness of the header jamb, plus ¼″
   B. double the header jamb thickness
   C. double the side jamb thickness, plus ½″
   D. ½″

_____ 4. Interior door frames are usually installed _____ .
   A. during the rough framing process
   B. after the interior wall covering is applied
   C. at the same time the interior wall covering is applied
   D. prior to the application of the interior wall covering

_____ 5. To find the width of the jamb stock, _____ .
   A. triple the thickness of the door
   B. subtract the thickness of the door jamb from that of the total wall thickness
   C. measure the total wall thickness including the wall covering
   D. subtract 1″ from the sill width

_____ 6. Door frames must be set so that the jambs are _____ .
   A. straight
   B. level
   C. plumb
   D. all of the above

_____ 7. If a rabbeted frame is used, _____ .
   A. the door's swing must be determined so that the rabbet faces the right direction
   B. a separate stop needs to be applied to the inside of the door frame
   C. the rough opening is the same width and height as the door frame
   D. the horns are not to be cut from the top edge of the side jambs

8. The header jamb is leveled by placing shims between the _____ .
   A. header jamb and the header
   B. header and the horns
   C. bottom of the appropriate side jamb and the subfloor
   D. side jamb opposite the header jamb

9. When straightening side jambs by shimming at intermediate points, use a _____ .
   A. combination square
   B. 6' straight edge
   C. butt gauge
   D. sliding T-bevel

10. Before any nails are driven home when setting a door frame, the frame should be checked for a _____ .
    A. wind
    B. bluster
    C. gale
    D. breeze

Completion

Complete each sentence by inserting the best answer on the line near the number.

1. Door stops are not permanently fastened until any necessary adjustment is made when the _____ is installed.

2. The special pivoting hardware installed on double acting doors returns the door to a _____ position after being opened.

3. Bypass door tracks are installed on the _____ according to the manufacturer's directions.

4. It is important that the door pulls on bypass doors are installed _____ so as not to obstruct the bypassing door.

5. Before installing bifold door tracks into position, be sure the _____ for the door pivot pins are inserted in the track.

6. Pocket door frames are usually assembled at the _____ .

7. _____ are tapered wood pieces used in pairs to adjust the space between materials.

8. To accommodate various wall thicknesses, prehung door units are available in various _____ widths.

9. To maintain proper clearance between the door and the frame on a prehung door, small cardboard _____ are stapled to the lock edge and the top end of the door.

10. The _____ lock is used often on bathroom and bedroom doors.

Math Problem-Solving

Solve the following math problems.

_____ 1. $a + b = b + ?$

_____ 2. $ab = b$ times ?

_____ 3. $(a + b) + c = ? + (b + c)$

_____ 4. $a(b + c) = ab + ?c$

_____ 5. $a \cdot 0 = ?$

_____ 6. $a + (-a) = ?$

_____ 7. $a \cdot 1 = ?$

_____ 8. Evaluate $3 - 8(4 + 5) \div (-3) =$

_____ 9. Evaluate $28 \div (-7)(2)^2 + 3(-4 - 2)^2 - (-3)^2$

_____ 10. If $x = 3$ and $y = 2$ evaluate $\dfrac{2(x + y) - 2y}{2(x - y)}$

78 Description and Application of Molding

Multiple Choice

Write the letter for the best answer on the line next to the number of the sentence.

_____ 1. Finger jointed molding is _____ .

    A. made from short pieces joined together
    B. designed to be painted
    C. installed the same as solid molding
    D. all of the above

_____ 2. Molding is classified by _____ .

    A. its size
    B. its shape and location of use
    C. the material it is made from
    D. all of the above

_____ 3. A molding used to conceal the joints between walls and ceiling is _____ .

    A. bed and crown
    B. casing and back band
    C. apron and stool
    D. quarter round and corner guard

_____ 4. A miter cut is made in one pass with the saw set _____ .

    A. at 45°
    B. at 90°
    C. to cut a compound angle
    D. to cut a double angle

_____ 5. When cutting ceiling molding in a miter box, position the side that will touch the ceiling _____ .

    A. upside down
    B. face up
    C. against the fence of the miter saw
    D. against the base of the miter saw

_____ 6. Miter cuts are made with a _____ .

    A. miter box
    B. table saw
    C. hand saw
    D. all the above

_____ 7. Coped joints are used for _____ .

    A. inside corners only
    B. outside corners only
    C. both inside and outside corners
    D. neither inside or outside corners

———— 8. Casing is applied around _____ .
   A. ceilings
   B. windows and doors
   C. walls about 3' above the floor
   D. all of the above

Completion

Complete each sentence by inserting the best answer on the line near the number.

_____ 1. In order to present a suitable appearance, moldings must be applied with _____ joints.

_____ 2. To reduce waste, door casings are available in lengths of _____ feet.

_____ 3. Finger-jointed lengths of molding should only be used when a _____ finish is to be applied.

_____ 4. Moldings are either classified by their _____ or their designated location.

_____ 5. The joint between the bottom of the base and the finish floor is usually concealed by the base _____ .

_____ 6. _____ are used to trim around windows, doors, and other openings to cover the space between the wall and the frame.

_____ 7. For a more decorative appearance, _____ may be applied to the outside edges of casings.

_____ 8. Aprons and stools are part of _____ trim.

_____ 9. Another name for outside corners is corner _____ .

_____ 10. Caps and chair rails are used to trim the top edge of _____ .

_____ 11. With the exception of prefinished molding, joints between molding lengths should be _____ flush after the molding has been fastened.

_____ 12. Molding joints on exterior corners must be _____ .

_____ 13. Joints on interior corners, especially on large moldings, are usually _____ .

_____ 14. When cutting molding in a miter box, the face side of the molding should be placed _____ .

_____ 15. When mitering bed, crown, and cove moldings, place them _____ in the miter box.

_____ 16. The _____ is a tool to use when paper-thin corrective cuts are needed when fitting miter joints.

17. A mitering _____ may be used to make miters on a table or radial arm saw quickly and easily without any change in the setup.

18. When making a coped joint, the coping saw's handle should be above the work and its teeth should face _____ from the saw's handle.

19. To assure straight application of large-size ceiling moldings, a _____ should be used as a guide.

20. Nails should be placed 2″ to 3″ from the molding's end to prevent _____ .

Identification: Molding

Identify each term, and write the letter of the best answer on the line next to each number.

_____ 1. crown

_____ 2. bed

_____ 3. cove

_____ 4. quarter round

_____ 5. corner guard

_____ 6. base shoes

_____ 7. chair rail

_____ 8. base moldings

_____ 9. half round

_____ 10. hand rail

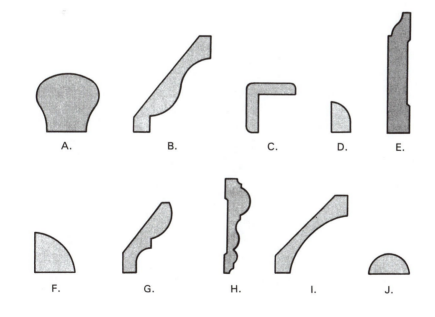

A. B. C. D. E.

F. G. H. I. J.

Math Problem-Solving

Simplify each statement below by removing parentheses.

_____ 1. $a - (b + c)$

_____ 2. $a - (-b - c)$

_____ 3. $(2x + 3) + (4y + 5z + 6)$

_____ 4. $(2x + 4) - (-2y + 2)$

_____ 5. $-(3x + y) - (2z + 7w) - (3r - 5s + 2)$

79 Application of Door Casings, Base, and Window Trim

Multiple Choice

Write the letter for the best answer on the line next to the number of the sentence.

_____ 1. Door casings must be applied before _____ .
   A. window trim
   B. ceiling moldings
   C. base moldings
   D. corner guards

_____ 2. _____ blocks are small decorative blocks used as part of the door trim at the base and at the head.
   A. Plinth
   B. Batten
   C. Astragal
   D. Crown

_____ 3. Molded casings usually have their back sides _____ .
   A. rabbeted
   B. backed out
   C. beveled
   D. laminated

_____ 4. An alternative to using molded casings is to use _____ stock.
   A. S4S
   B. S3S
   C. S2S
   D. S1S

_____ 5. The ³⁄₁₆″ to ⁵⁄₁₆″ setback of door casings from the inside face of the door frame is called the _____ .
   A. reveal
   B. exposure
   C. rake
   D. salvage

_____ 6. For each interior door opening, _____ are required.
   A. two side casings and two header casings
   B. two side casings and four header casings
   C. four side casings and four header casings
   D. four side casings and two header casings

_____ 7. When fastening casing into _____ , use 6d or 8d finish nails.
   A. header jambs
   B. side jambs
   C. framing
   D. all of the above

_____ 8. To bring the faces of mitered casing joints flush, _____ .
   A. sand them
   B. shim between the casing back and the wall
   C. use a wood chisel to remove thin shavings from the thicker side of the joint
   D. use a flat wood file

_____ 9. Base trim should be _____ .
   A. thinner than the door casing
   B. molded or S2S stock
   C. back mitered to outline the cope for interior corners
   D. all of the above

_____ 10. When fastening base molding, use _____ finishing nail(s) of sufficient length at each stud location.
   A. one
   B. two
   C. three
   D. four

Completion

Complete each sentence by inserting the best answer on the line near the number.

_____ 1. When scribing lines on baseboard, be sure to hold the dividers so that a line between the two points is _____ to the floor.

_____ 2. When cutting baseboard to a scribed line, be sure to give the cut a slight _____ .

_____ 3. On outside corners of baseboard, make regular _____ joints.

_____ 4. The base shoe is usually nailed into the _____ .

_____ 5. No base shoe is required if _____ is used as a finish floor.

_____ 6. The bottom side of the stool is _____ at an angle, so its top side will be level after it is fit on the window's sill.

_____ 7. The _____ covers the joint between the sill and the wall.

_____ 8. Jamb _____ must be used when windows are installed with jambs that are narrower than the walls.

_____ 9. Window casings are installed with their inside edges flush with the inside face of the _____ .

_____ 10. The _____ is a piece of 1″ × 5″ stock installed around the walls of a closet to support the rod and the shelf.

_____ 11. Closet pole sockets should be located at least _____ inches from the back wall.

_____ 12. When sanding interior trim, always sand with the _____ .

_____ 13. All traces of excess glue must be removed if the trim is to be _____ .

_____ 14. To help prevent the hammer from glancing off the head of a nail, occasionally clean the hammer's face by rubbing it with _____ .

Math Problem-Solving

Solve the following math problems.

_____ 1. How many lineal feet of baseboard should be estimated for a room that measures 12′ × 18′ if the room has two 3′ doorways?

_____ 2. How many lineal feet of crown molding should be estimated for a room that measures 18′ × 22′ and has two 3′ doorways?

_____ 3. How many lineal feet of casing should be estimated for three 2′-6″ × 6′-8″?

_____ 4. How many lineal feet of casing should be estimated for a 3656 window that is cased on four sides?

_____ 5. How many lineal feet of baseboard should be estimated for a group of interior partitions that total 120′ in length and have five 3′ doorways?

80 Description of Stair Finish

Multiple Choice

Write the letter for the best answer on the line next to the number of the sentence.

_____ 1. The _____ is a component of the stair body finish.
   A. newel post
   B. riser
   C. handrail
   D. baluster

_____ 2. With an open staircase, _____ .
   A. tread ends butt against the wall
   B. one or more ends of the treads are exposed to view
   C. the area under the stair body is exposed to view
   D. both sides of the treads must be exposed to view

_____ 3. The part of the stair tread that extends beyond the riser is the _____ .
   A. nosing
   B. tread molding
   C. stringer
   D. half newel

_____ 4. Finish stringers are sometimes called _____ .
   A. preachers
   B. skirt boards
   C. carriages
   D. aprons

_____ 5. Open finish stringers are placed on the _____ .
   A. open side of the stairway above the treads
   B. closed side of an open stairway above the treads
   C. open side of a stairway below the treads
   D. closed side of an open stairway below the treads

_____ 6. When the staircase is open on one or more sides, _____ .
   A. a starting step may be used
   B. return nosing is used
   C. an open finish stringer is used
   D. all of the above

_____ 7. In post to post balustrades _____ .
   A. the newel posts fit into the bottom of the handrail
   B. the newel posts have flat square surfaces near the top
   C. goosenecks must always be used
   D. the newel posts are made with a pin in their tops

_____ 8. Narrow strips called _____ are used between balusters to fill the plowed groove on handrails and shoe rails.
  A. fillets
  B. rossettes
  C. mullions
  D. scabs

_____ 9. Short sections of specialty curved handrail are called _____ .
  A. stiles
  B. fittings
  C. heels
  D. crickets

_____ 10. _____ are vertical, usually decorative, pieces between newel posts and spaced close together, supporting the handrail.
  A. Volutes
  B. Cleats
  C. Battens
  D. Balusters

Identification: Balustrade Components

Identify each term, and write the letter of the correct answer on the line next to each number.

_____ 1. baluster

_____ 2. rake handrail

_____ 3. landing newel

_____ 4. closed finish stringer

_____ 5. half newel

_____ 6. starting newel

_____ 7. landing rail

_____ 8. balcony handrail

_____ 9. landing baluster

_____ 10. balcony baluster

Math Problem-Solving

Combine like terms.

_____ 1. $4b + b$

_____ 2. $2x - 3y + 3x + 4y$

_____ 3. $x^2 + 2x^2 + 3x + 6x$

_____ 4. Add the terms
$$\begin{array}{r} +4x^2 - 7x + 4 \\ -2x^2 + 4x - 5 \end{array}$$

_____ 5. $(2x^2 - 5x - 2) - (x^2 - x + 8)$

81 Finishing Open and Closed Staircases

Multiple Choice

Write the letter for the best answer on the line next to the number of the sentence.

_____ 1. Use a _____ when cutting the level and plumb cuts on a closed finish stringer.
   A. 7- or 8-point crosscut handsaw
   B. fine tooth crosscut handsaw
   C. 5½-point hand ripsaw
   D. coping saw

_____ 2. The nosed edge of the tread usually extends beyond the face of the riser by _____ .
   A. ⅝"
   B. ⅞"
   C. 1⅛"
   D. 1⅝"

_____ 3. Prior to placing the treads between the finish stringers, _____ .
   A. rub wax on one end
   B. liberally apply glue to the tread's end grain
   C. spray the treads with a light mist of water
   D. seal the treads with a varnish

_____ 4. When fastening treads, to prevent splitting the wood, _____ .
   A. predrill the nail holes
   B. use only 12d finish nails
   C. be sure to use galvanized 10d casing nails
   D. use only glue and no nails

_____ 5. Tread molding is attached using _____ finish nails.
   A. 3d
   B. 4d
   C. 6d
   D. 8d

_____ 6. On a staircase that has one side open and the other side closed, the _____ the first item(s) of finish to be applied.
   A. risers are
   B. closed finish stringer is
   C. open finish stringer is
   D. treads are

_____ 7. When laying out the plumb lines on an open finish stringer, a helpful device to use is a _____ .
   A. chalk line
   B. plumb bob
   C. line level
   D. preacher

_____ 8. When fastening a mitered riser to a mitered stringer, _____ .
   A. sand all pieces before installation
   B. apply a small amount of glue to the joint
   C. drive finish nails both ways through the miter
   D. all of the above

_____ 9. When ripping treads to width, _____ .
        A. bevel the back edge
        B. make allowances for the rabbeted back edge
        C. bevel the front edge of the tread
        D. only a hand ripsaw should be used

_____ 10. Treads on winding steps _____ .
        A. are more challenging to fit
        B. have the same angle on both ends
        C. are wider on the inside than the outside
        D. all of the above

Completion

Complete each sentence by inserting the best answer on the line near the number.

_____ 1. Instead of a framing square, a _____ may be used to lay out a housed stringer.

_____ 2. The _____ is joined to the housed stringer at the top and bottom of the staircase.

_____ 3. A _____ is used to guide the router when housing the stringer.

_____ 4. Housed stringers are routed so that the dadoes are the exact width at the nosing and wider toward the inside so the treads and risers can be _____ against the shoulders of the dadoes.

_____ 5. An open stringer is also referred to as a _____ stringer.

_____ 6. The vertical layout line on an open stringer is mitered to fit the mitered end of the _____ .

_____ 7. On open stringers, a _____ cut is made through the stringer's thickness for the tread.

_____ 8. A ⅜″ by ⅜″ groove to receive the rabbeted inner edge of the _____ may be cut on the face side of all but the first riser.

_____ 9. If the staircase is open, a return _____ is mitered to the end of the tread.

_____ 10. _____ are usually the first members applied to the stair carriage in a closed staircase.

_____ 11. The top edge of a closed finished stringer is usually about _____ inches above the tread nosing.

_____ 12. When cutting along the layout lines for a closed finished stringer, extreme care must be made on the _____ cut.

Math Problem-Solving

Simplify the following expressions.

_____ 1. $(5a)(-3)$

_____ 2. $(6x^2)(4x)$

_____ 3. $(6a^2b)(8a)$

_____ 4. $(y^2)^3$

_____ 5. If $a = 3$ and $b = -2$, solve $(-2a^2)(3ab)$

82 Balustrade Installation

Multiple Choice

Write the letter for the best answer on the line next to the number of the sentence.

_____ 1. The first step in laying out balustrades is to _____ .
   A. lay out the handrail
   B. determine the rake of the handrail
   C. lay out the balustrade's centerline
   D. determine the height of the starting newel

_____ 2. Most building codes require that balusters be spaced so that _____ .
   A. a small child's foot cannot fit between them
   B. there is no more than one per tread
   C. the space between them equals their width
   D. no object 4" in diameter can pass through

_____ 3. On open treads, the center of the front baluster is located _____ .
   A. a distance equal to ½ its thickness back from the riser's face
   B. flush with the riser's face
   C. on the center of the tread's outside edge
   D. 2" from the riser face

_____ 4. Rake handrail height is the vertical distance from the tread nosing to the _____ .
   A. handrail bottom side
   B. handrail center
   C. top of the handrail
   D. top of the newel post

_____ 5. On a stairway more than 88" in width, a handrail _____ .
   A. is needed on only one side
   B. is needed on both sides
   C. must be provided in the stairway's center
   D. B and C

_____ 6. The starting newel post is notched over the _____ .
   A. inside corner of the second step
   B. inside corner of the first step
   C. outside corner of the first step
   D. outside corner of the second step

_____ 7. Generally, codes require that balcony rails for homes be not less than _____ in length.
   A. 18"
   B. 24"
   C. 36"
   D. 42"

_____ 8. To determine the height of a balcony newel, you must know the _____ .
   A. height of the balcony handrail plus 1" for the block reveal
   B. height of the newel's turned top
   C. distance the newel extends below the floor
   D. all of the above

_____ 9. When square top balusters are used, _____ .

    A. holes must be bored in the bottom side of the handrail at least ¾" deep

    B. the balusters must be trimmed to length at the rake angle

    C. holes need not be bored in the treads

    D. all of the above

_____ 10. Instead of a half newel, a _____ is sometimes used to end the balcony handrail.

    A. landing fitting

    B. gooseneck

    C. half baluster

    D. rosette

Completion

Complete each sentence by inserting the best answer on the line near the number.

_____ 1. Square top balusters are fastened to the handrail with finish nails and _____ .

_____ 2. When using square top balusters, _____ are installed between the balusters in the plow of the handrail.

_____ 3. When installing an over-the-post balustrade, the first step is to lay out the balustrade and baluster _____ on the stair treads.

_____ 4. A _____ is a piece of wood cut in the shape of a right triangle, whose sides are equal in length to the rise and tread run of the stairs.

_____ 5. When laying out the starting fitting, mark it at the _____ point, where its curve touches the pitch block.

_____ 6. When cutting handrail fittings on a power miter box, be sure to securely _____ the fitting and the pitch block.

_____ 7. To mark the hole locations for handrail bolts, a _____ should be used to assure proper alignment.

_____ 8. A one-riser balcony gooseneck fitting is used when the balcony rails are _____ inches high.

_____ 9. On over-the-post balustrades, the height of the rake handrail is calculated from the height of the starting _____ .

_____ 10. The height of a balcony newel on an over-the-post balustrade is found by subtracting the handrail thickness from the handrail _____ .

Math Problem-Solving

Simplify the following expressions by removing the parentheses.

_____ 1. $5(a - 6)$ _____ 3. $a(4x^2 - 5y + 2)$ _____ 5. $(x - 3)(x - 5)$

_____ 2. $-3(6x - 4y^2)$ _____ 4. $(x + 1)(x + 6)$

83 Description of Wood Finish Floors

Multiple Choice

Write the letter for the best answer on the line next to the number of the sentence.

_____ 1. Most hardwood finish flooring is made from _____ .
- A. Douglas fir
- B. white or red oak
- C. hemlock
- D. southern yellow pine

_____ 2. The most widely used type of solid wood flooring is _____ .
- A. strip
- B. laminated parquet blocks
- C. laminated strip
- D. parquet strip

_____ 3. Unfinished strip flooring _____ .
- A. has a chamfer machined between the face and edge sides
- B. cannot be sanded after installation
- C. is milled with square sharp corners at the intersection of the face and the edges
- D. is waxed at the factory

_____ 4. Laminated strip flooring _____ .
- A. is easily recognizable by the V-grooves on the floor's surface after it is laid
- B. is a 5-ply prefinished wood assembly
- C. must be sanded after it is installed
- D. is chamfered on its edges

_____ 5. Plank flooring is similar to _____ flooring.
- A. strip
- B. parquet strip
- C. Monticello
- D. Marie Antoinette

_____ 6. The highest quality parquet block flooring is made with _____-thick, tongue and groove solid hardwood flooring.
- A. ⅜″
- B. ½″
- C. ⅝″
- D. ¾″

_____ 7. Monticello is the name of a parquet originally designed by _____ .
- A. Benjamin Franklin
- B. George Washington
- C. Thomas Jefferson
- D. Samuel Adams

_____ 8. Laminated blocks are generally made in a _____ thickness.
- A. ⅜″
- B. ½″
- C. ⅝″
- D. ¾″

_____ 9. Finger blocks are another name for _____ .

    A. laminated blocks
    B. slat blocks
    C. unit blocks
    D. all of the above

_____ 10. _____ is the top grade of unfinished oak flooring.

    A. No. 1
    B. No. 2
    C. Clear
    D. Select

Completion

Complete each sentence by inserting the best answer on the line near the number.

_____ 1. In addition to appearance, grades are based on _____ .

_____ 2. Red grades of unfinished pecan flooring contain all _____ .

_____ 3. White grades of unfinished pecan flooring contain all _____ .

_____ 4. The lowest grade of unfinished hard maple flooring is called _____ grade.

_____ 5. The average length of clear bundles of flooring is _____ feet.

_____ 6. A bundle of flooring may contain pieces from 6″ under to 6″ over the _____ length of the bundle.

_____ 7. No pieces shorter than _____ inches in length are allowed in a bundle.

_____ 8. _____ is the lowest grade of prefinished flooring.

Math Problem-Solving

Solve the following math problems using information for the room shown in the figure.

_____ 1. What is the longest dimension of the room?

_____ 2. What is the area of the room?

_____ 3. How many board feet of wood strip flooring should be estimated for 1½″ wide boards?

_____ 4. How many board feet of flooring should be estimated if 2¼″ boards are used? Round up answer to the nearest board foot.

_____ 5. If the entire room is to have a boarder strip installed, how many feet of boarder should be estimated?

12′
10′
8′
14′

84 Laying Wood Finish Floor

Multiple Choice

Write the letter for the best answer on the line next to the number of the sentence.

_____ 1. Before wood strip flooring is installed it is recommended that the _____ .
A. subfloor be ½″ thick
B. wood acclimate for several days
C. floor be insulated
D. all of the above

_____ 2. When laying the first course of strip flooring, it is placed with its _____ from the starting wall.
A. tongue side ¾″
B. groove side ½″
C. tongue side ½″
D. groove side ¾″

_____ 3. When blind nailing flooring, drive the nails at about a _____ angle.
A. 90°
B. 60°
C. 45°
D. 30°

_____ 4. When nailing strip flooring by hand, _____ is used to set hardened flooring nails.
A. the head of the next nail to be driven
B. a nail set
C. a screwdriver laid on its side
D. a center punch

_____ 5. Laying out loose flooring ahead of time to assure efficient installation is known as _____ .
A. cribbing the stock
B. racking the floor
C. grouping the strips
D. clustering the stock

_____ 6. When using a power nailer to fasten flooring, _____ .
A. an air compressor must be used
B. holes for the fasteners must be predrilled
C. one blow must be used to drive the fastener
D. the floor layer must not stand on the flooring

_____ 7. The last course of strip flooring must be _____ .
A. fastened with a power nailer
B. blind nailed
C. face nailed
D. glued and not nailed

_____ 8. Laminated strip flooring _____ .
   A. needs an ⅛″ foam underlayment
   B. is not fastened or cemented to the floor
   C. is brought up tight against the previous course with a hammer and tapping block
   D. all of the above

_____ 9. When installing parquet flooring, it is usually _____ .
   A. face nailed
   B. blind nailed
   C. laid in mastic
   D. fastened with a power nailer

_____ 10. When laying unit blocks in a square pattern, two layout lines are snapped _____ .
   A. at right angles to each other and diagonal to the walls
   B. with one parallel and one diagonal to the walls
   C. at right angles to each other parallel to the walls
   D. parallel to each other and diagonal to the walls

Completion

Complete each sentence by inserting the best answer on the line near the number.

_____ 1. The installation of wood finish floors on _____ slabs is not recommended.

_____ 2. New concrete slabs should be allowed to age at least _____ days prior to the installation of a wood finish floor.

_____ 3. When preforming a rubber mat moisture test on a concrete slab, allow the mat to remain in place at least _____ hours.

_____ 4. To ensure a trouble-free finish floor installation, a _____ must be installed over all concrete slabs.

_____ 5. Prior to the application of polyethylene film to a concrete slab, a skim coat of _____ is troweled over the entire area.

_____ 6. Exterior grade sheathing plywood may be used for a subfloor if it is at least _____ inch(es) thick.

_____ 7. Strips of wood laid over a concrete floor that finish flooring is attached to are known as _____ .

_____ 8. The National Oak Flooring Manufacturers Association does not recommend fastening finish flooring to subfloors of _____ panels.

_____ 9. For its best appearance, strip flooring should be laid in the direction of the room's _____ dimension.

_____ 10. To help keep out dust, prevent squeaks, and retard moisture from below, _____ is applied over the subflooring.

Math Problem-Solving

Simplify each equation for the unknown.

_____ 1. $3a = 18$

_____ 2. $\dfrac{2}{3}n = 6$

_____ 3. $n + 12 = -5$

_____ 4. $4x + 9 = 7x - 15$

_____ 5. $-3x + 17 = 6x - 37$

_____ 6. $17 - 4y = 14 - y$

_____ 7. $2(x + 3) - 6 = 10$

_____ 8. $-3x + 5(x - 6) = 32$

_____ 9. $8x - 4(x + 4) - 12 = 0$

_____ 10. $16 = -3(x - 4)$

Discussion

Write your answer(s) on the lines below.

1. Describe the recommended procedures that are necessary to maintain the proper moisture content of hardwood flooring prior to its installation.

   _____

   _____

   _____

85 Underlayment and Resilient Tile

Multiple Choice

Write the letter for the best answer on the line next to the number of the sentence.

_____ 1. Resilient flooring is installed _____ .
   A. with construction adhesive
   B. on underlayment
   C. after baseboard is applied
   D. all of the above

_____ 2. Before resilient tile is installed _____ .
   A. the floor should be swept clean
   B. underlayment nails must be properly set
   C. large spaces in floor must be filled
   D. all of the above

_____ 3. Underlayment is installed with _____ .
   A. fasteners that penetrate the floor joists
   B. its grain perpendicular to the subfloor
   C. edges that do not align with the subfloor
   D. all of the above

_____ 4. The tool used to apply resilient floor tiles is _____ .
   A. putty knife
   B. notched trowel
   C. brush or roller
   D. hands and fingers

_____ 5. Resilient floor tile is ready for installation when the adhesive _____ .
   A. turns lighter in color
   B. is totally dry
   C. has been on the floor for several hours
   D. becomes transparent and dry to the touch

_____ 6. Border tiles are installed _____ .
   A. along the walls
   B. around door openings
   C. around the perimeter of the room
   D. all of the above

_____ 7. Often the grain of the tiles is installed _____ .
   A. randomly
   B. in the alternate directions
   C. to the left
   D. to the right

_____ 8. When installing flooring tiles take care to _____ .
   A. keep a ¹⁄₁₆″ gap between them
   B. slide them into position
   C. lay them in place without sliding them
   D. lay border tiles first

_____ 9. The perimeter of a room is found by _____ .
   A. adding twice the width to twice the length
   B. adding the length to the width
   C. multiplying the length times the height
   D. multiplying the length times the width

_____ 10. The number of 12″ × 12″ floor tiles needed for a rectangular room that measures 12'-3″ × 26'-3″ is _____ tiles.
   A. 312
   B. 322
   C. 324
   D. 351

Completion

Complete each sentence by inserting the best answer on the line near the number.

_____ 1. _____ is installed on top of the subfloor to provide a base for the application of resilient sheet or tile flooring.

_____ 2. All joints between the subfloor and the underlayment should be _____ .

_____ 3. To allow for expansion, leave about _____ inch(es) between underlayment panels.

_____ 4. Underlayment face grain should run _____ the floor joists.

_____ 5. If _____ is used as a subfloor, no underlayment is needed.

_____ 6. Resilient floor tiles are applied to the floor in a manner similar to applying _____ .

_____ 7. Long strips called _____ strips are used between the tiles to create unique floor patterns.

_____ 8. Caution must be exercised when removing existing resilient floor covering because it may contain _____ .

_____ 9. Before installing resilient tile, make sure underlayment _____ are not projecting above the surface.

_____ 10. Resilient tile adhesive may be applied over the _____ area before tiles are installed.

_____ 11. Cutting curves in resilient tile is made easier by using a _____ .

_____ 12. When applying adhesive for floor tile, it is important that the trowel has the proper size _____ .

_____ 13. When laying tiles, start at the _____ of the layout lines.

_____ 14. Tiles are to be laid in place instead of _____ them into position.

_____ 15. Border tiles may be cut by scoring with a sharp _____ and bending.

_____ 16. Many times a vinyl _____ is used to trim a tile floor.

_____ 17. The number of 12″ × 12″ tiles needed to cover a floor is equal to the floor's _____ in square feet.

Math Problem-Solving

Simplify the following math problems using information of a room shown in the figure. Round each answer up to a whole number.

_____ 1. What is the area of the room?

_____ 2. How many pieces of 12″ × 12″ tile should be ordered if 7% is added for waste?

_____ 3. If the resilient floor tile was to be centered in the larger rectangular area of the room, what would be the two dimensions of the corner boarder tile?

_____ 4. What is the perimeter of the room?

_____ 5. If the boarder tile was to be a different color from the main floor, how many 12″ × 12″ tiles should be estimated for the entire perimeter?

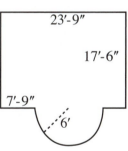

Discussion

Write your answer(s) on the lines below.

1. Unless we know for sure that it does not, why should we assume that existing flooring contains asbestos?

   _____

   _____

   _____

86 Description and Installation of Manufactured Cabinets

Multiple Choice

Write the letter for the best answer on the line next to the number of the sentence.

_____ 1. The method of cabinet construction that has a traditional look is called _____ .
   A. face-framed
   B. European
   C. modular
   D. prosaic

_____ 2. The two basic kinds of kitchen cabinet units are _____ .
   A. supporting and shelf
   B. standing and attached
   C. base and wall
   D. case and drawer

_____ 3. Countertops are usually_____ from the floor.
   A. 28″
   B. 32″
   C. 36″
   D. 40″

_____ 4. The usual overall height of a kitchen cabinet installation is _____ .
   A. 6′-6″
   B. 7′-0″
   C. 7′-6″
   D. 8′-0″

_____ 5. Standard wall cabinets are _____ deep.
   A. 8″
   B. 10″
   C. 12″
   D. 14″

_____ 6. The usual countertop thickness is _____ .
   A. ¾″
   B. 1″
   C. 1¼″
   D. 1½″

_____ 7. A recess called a _____ is provided at the bottom of the base cabinet.
   A. toe kick
   B. foot area
   C. counter base
   D. floor cove

_____ 8. Cabinets that provide access from both sides are called _____ .
   A. dual-sided
   B. twin-entry
   C. bi-frontal
   D. double-faced

_____ 9. Double door pantry cabinets are made _____ wide.
   A. 28″
   B. 36″
   C. 42″
   D. 48″

_____ 10. Wall cabinets with a 24″ depth are usually installed _____ .
   A. when a contemporary appearance is desired
   B. above refrigerators and tall cabinets
   C. above ranges
   D. all of the above

Completion

Complete each sentence by inserting the best answer on the line near the number.

_____ 1. Most vanity base cabinets are manufactured _____ inches high.

_____ 2. The first step in drawing a cabinet layout plan is to carefully and accurately _____ the walls on which the cabinets are to be installed.

_____ 3. Many large kitchen cabinet distributors will, on request, provide _____ .

_____ 4. Scribbing and fitting cabinets to an uneven floor eliminates the need for a _____ .

_____ 5. When laying out the wall, a level line is drawn _____ up the wall to indicate the top of the base cabinets.

_____ 6. When installing cabinets, most installers prefer to mount the _____ units first.

_____ 7. The installation of wall cabinets is started in a _____ .

_____ 8. If base cabinets are to be fitted to the floor, then their level layout line is measured from the _____ of the floor.

_____ 9. Countertops are covered with a thin, tough, high-pressure plastic surface known as _____ .

_____ 10. To prevent scratching the countertop when making the sink cutout, apply _____ to the base of the saber saw.

Math Problem-Solving

Solve each for the unknown indicated.

_____ 1. $\dfrac{2x}{3} = \dfrac{32}{6}$ _____ 3. $\dfrac{3}{8}x = \dfrac{14}{16}$ _____ 5. $\dfrac{3}{x} - 8 = 7$

_____ 2. $\dfrac{5x}{7} = \dfrac{20}{14}$ _____ 4. $\dfrac{4}{x} = 6$

87 Countertops and Cabinet Components

Multiple Choice

Write the letter for the best answer on the line next to the number of the sentence.

_____ 1. Wrought iron and other decorative hinges are usually _____ hinges.
   A. pivot
   B. concealed offset
   C. European-style
   D. surface

_____ 2. Offset hinges are used on _____ doors.
   A. overlay
   B. flush
   C. lipped
   D. European-style

_____ 3. Pivot hinges are usually used on _____ doors.
   A. overlay
   B. flush
   C. lipped
   D. European-style

_____ 4. Many carpenters use a self-centering tool, called a _____ , when drilling pilot holes for a screw fastening of cabinet door hinges.
   A. VIX bit
   B. expansive bit
   C. butt gauge
   D. butt marker

_____ 5. When two screws are used to fasten a pull, drill the holes _____ the diameter of the screw.
   A. smaller than
   B. the same size as
   C. slightly larger than
   D. twice

Completion

Complete each sentence by inserting the best answer on the line near the number.

_____ 1. Most countertops are covered with _____ .

_____ 2. Before laminating a countertop, lightly hand or power sand all joints, making sure they are _____ .

_____ 3. When trimming laminate with a router or a laminate trimmer, it is recommended to use a _____ trimming bit.

_____ 4. _____ cement must be dry before the laminate is bonded to the core.

_____ 5. To test the cement for dryness, you can check it with your _____ .

_____ 6. If the cement is allowed to dry more than _____ hours, the laminate will not bond properly.

_____ 7. When using a trimming bit with a dead pilot, the laminate must be _____ where the pilot will ride.

_____ 8. To prevent water from seeping between the backsplash and the countertop, apply _____ to the joint.

_____ 9. If heated to 325°F, the laminate can be bent to a minimum radius of _____ inches.

_____ 10. The most widely used method of hanging cabinet doors is the _____ .

_____ 11. Face frames are not used on _____-style cabinets.

_____ 12. The _____ door has rabbeted edges that overlap the opening by about ⅜″ on all sides.

_____ 13. The _____-type door must be fitted to the door opening.

_____ 14. Drawer fronts are generally made from the same material as the cabinet _____ .

_____ 15. Drawer sides and backs are usually _____ inch thick.

_____ 16. Drawer bottoms are usually made of _____-inch-thick plywood or particleboard.

_____ 17. The _____ joint is the strongest used in drawer construction.

_____ 18. The _____ joint is the easiest joint to make that is used between the drawer front and side.

_____ 19. To provide added strength, the drawer back is usually set _____ inch(es) from the rear of the sides.

_____ 20. The simplest type of wood drawer guide is probably the _____ .

Identification: Wood Joints

Identify each term, and write the letter of the best answer on the line next to each number.

_____ 1. dovetail joint

_____ 2. dado joint

_____ 3. dado and rabbet joint

_____ 4. butt joint

_____ 5. lock joint

_____ 6. rabbeted joint

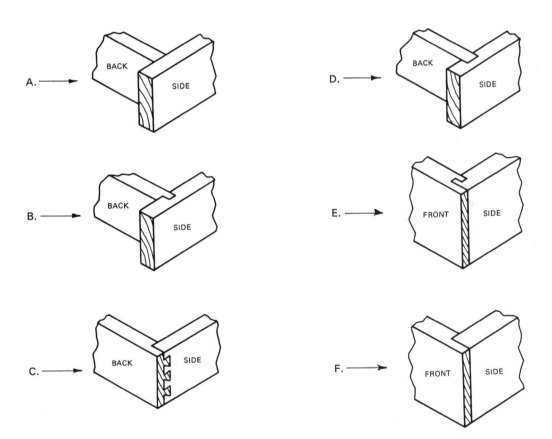

Math Problem-Solving

Solve each for the unknown indicated.

_____ 1. $E = Ir$ for r

_____ 2. $F = ma$ for a

_____ 3. $E = mc^2$ for m

_____ 4. $v^2 = 2gh$ for h

_____ 5. $F = \dfrac{9}{5}C + 32$ for C

Section 4: Building for Success

DEMONSTRATING QUALITY WORKMANSHIP

The demonstration of quality workmanship begins with two basic premises: one involving need and one involving desire. First, each skilled worker must accept and understand the need to do quality work. Skill competencies cannot be reached without a commitment to quality. Second, each person has to want to do quality work. Without meeting these two requirements, quality workmanship cannot be accomplished.

Because customers expect quality materials and workmanship and are willing to pay for them, they will not settle for anything less. By looking at successful building and remodeling companies, quality work will surface as a trademark. With those companies quality will become a product. Therefore, all workers associated with those companies are expected to deliver quality. The mind-set for each person has to be a total buy-in to the quality concept.

How is quality to be defined? What are the characteristics of quality? In some cases, definitions of quality work will differ. Its definition is often based on expectations. At the beginning of each construction project, written statements of quality in materials, features, and workmanship must relate to expectations held by all parties. This in turn relates to satisfaction. Therefore it becomes imperative to agree on quality materials and workmanship before the construction process begins.

There are quality standard guidelines available for both new construction and remodeling. For example, *The Quality Standards for the Professional Remodeler,* published by the NAHB Remodeler's Council, has established good descriptions of industry quality standards. Other ideal publications by the NAHB Home Builder Bookstore are: *Production Checklist for Builders and Superintendents, Customer Service for Home Builders, The Positive Walk Through: Your Blueprint for Success,* and *Warranty Service for Builders and Remodelers.*

Typically, these standards manuals and guides present criteria, written specifications, and suggestions that help each builder or remodeler provide quality products, workmanship, and service. The quality standards presented in these publications define measureable criteria to be used throughout the construction process. These manuals serve well in resolving complaints and disputes between the builder or remodeler and the client. They are written in layperson's terms that are easily understood by everyone.

Written materials that explain quality workmanship and construction procedures set the stage for credibility in relationships in a construction project. Written standards that are presented at the beginning of the construction discussions reassure clients that they will receive the quality they are paying for. With that assurance, clients will be more trusting and willing to cooperate fully with the builder or remodeler.

Quality standards, along with construction building codes, can be used by the builder or remodeler as the very minimum in quality to be provided throughout the building process. The goal should be to meet or exceed these minimums during the entire construction process.

The minimum quality standards adopted by a company also become the basic guidelines for training employees in company expectations. Each prospective employee should be exposed to these quality standards before actual employment. This will give everyone a chance to see what is expected on the job. At that time the potential employee can see the company philosophy, the commitment to quality workmanship, and how it is provided to the customer. This preview of expectations will undoubtedly increase each employee's chances for success.

By involving all employees in developing and using quality standards, a company can better promote and defend its commitment to excellence. Each skilled craftsperson should desire to protect his or her image of doing quality work. Communicating in writing what quality workmanship is and what each client can expect from the skilled workers should be presented in the early stages of the project. The commitment by the builder or remodeler to meet or exceed those standards will be satisfying to the client.

Competition will never allow the skilled worker to abandon the need to demonstrate quality workmanship. Each worker must develop good technical skills and gain a good understanding of the construction processes used in building. Obtaining and maintaining quality tools will help install quality materials. People are willing to pay for these things and more if they are satisfied the results will be the best possible. Customer satisfaction will remain one of the primary gauges for determining the level of quality workmanship and service being provided by a construction company. The margin for error is small and the tolerance level for poor workmanship will continue to decrease. Each person entering the construction industry today will have to buy into the quality workmanship concept to be successful.

Focus Questions:
For individual or group discussion

1. What criteria are necessary in order for a carpenter to demonstrate quality workmanship? Explain how these criteria will be observable to the client.
2. Is there a correlation between a carpenter's attitude toward quality workmanship and customer satisfaction? Explain your answer in detail.
3. What would be some advantages of starting the degree of quality materials and workmanship before a contract is agreed upon?
4. How could meeting or exceeding the quality standards and workmanship stated in the agreement prove to be a positive long-term strategy for the builder? Explain how this might affect customer satisfaction.